T0073822

Introduction to Algorithms for Data Mining and Machine Learning

Introduction to Algorithms for Data Mining and Machine Learning

Xin-She Yang
Middlesex University
School of Science and Technology
London, United Kingdom

ACADEMIC PRESS
An imprint of Elsevier

Academic Press is an imprint of Elsevier
125 London Wall, London EC2Y 5AS, United Kingdom
525 B Street, Suite 1650, San Diego, CA 92101, United States
50 Hampshire Street, 5th Floor, Cambridge, MA 02139, United States
The Boulevard, Langford Lane, Kidlington, Oxford OX5 1GB, United Kingdom

Notices

Knowledge and best practice in this field are constantly changing. As new research and experience broaden our understanding, changes in research methods, professional practices, or medical treatment may become necessary.

Practitioners and researchers must always rely on their own experience and knowledge in evaluating and using any information, methods, compounds, or experiments described herein. In using such information or methods they should be mindful of their own safety and the safety of others, including parties for whom they have a professional responsibility.

To the fullest extent of the law, neither the Publisher nor the authors, contributors, or editors, assume any liability for any injury and/or damage to persons or property as a matter of products liability, negligence or otherwise, or from any use or operation of any methods, products, instructions, or ideas contained in the material herein.

Library of Congress Cataloging-in-Publication Data
A catalog record for this book is available from the Library of Congress

British Library Cataloguing-in-Publication Data
A catalogue record for this book is available from the British Library

ISBN: 978-0-12-817216-2

For information on all Academic Press publications
visit our website at https://www.elsevier.com/books-and-journals

Publisher: Candice Janco
Acquisition Editor: J. Scott Bentley
Editorial Project Manager: Michael Lutz
Production Project Manager: Nilesh Kumar Shah
Designer: Miles Hitchen

Typeset by VTeX

Contents

About the author

Xin-She Yang obtained his PhD in Applied Mathematics from the University of Oxford. He then worked at Cambridge University and National Physical Laboratory (UK) as a Senior Research Scientist. Now he is Reader at Middlesex University London, and an elected Bye-Fellow at Cambridge University.

He is also the IEEE Computer Intelligence Society (CIS) Chair for the Task Force on Business Intelligence and Knowledge Management, Director of the International Consortium for Optimization and Modelling in Science and Industry (iCOMSI), and an Editor of Springer's Book Series *Springer Tracts in Nature-Inspired Computing* (STNIC).

With more than 20 years of research and teaching experience, he has authored 10 books and edited more than 15 books. He published more than 200 research papers in international peer-reviewed journals and conference proceedings with more than 36 800 citations. He has been on the prestigious lists of Clarivate Analytics and Web of Science highly cited researchers in 2016, 2017, and 2018. He serves on the Editorial Boards of many international journals including *International Journal of Bio-Inspired Computation*, Elsevier's *Journal of Computational Science (JoCS)*, *International Journal of Parallel, Emergent and Distributed Systems*, and *International Journal of Computer Mathematics*. He is also the Editor-in-Chief of the *International Journal of Mathematical Modelling and Numerical Optimisation*.

Preface

Both data mining and machine learning are becoming popular subjects for university courses and industrial applications. This popularity is partly driven by the Internet and social media because they generate a huge amount of data every day, and the understanding of such big data requires sophisticated data mining techniques. In addition, many applications such as facial recognition and robotics have extensively used machine learning algorithms, leading to the increasing popularity of artificial intelligence. From a more general perspective, both data mining and machine learning are closely related to optimization. After all, in many applications, we have to minimize costs, errors, energy consumption, and environment impact and to maximize sustainability, productivity, and efficiency. Many problems in data mining and machine learning are usually formulated as optimization problems so that they can be solved by optimization algorithms. Therefore, optimization techniques are closely related to many techniques in data mining and machine learning.

Courses on data mining, machine learning, and optimization are often compulsory for students, studying computer science, management science, engineering design, operations research, data science, finance, and economics. All students have to develop a certain level of data modeling skills so that they can process and interpret data for classification, clustering, curve-fitting, and predictions. They should also be familiar with machine learning techniques that are closely related to data mining so as to carry out problem solving in many real-world applications. This book provides an introduction to all the major topics for such courses, covering the essential ideas of all key algorithms and techniques for data mining, machine learning, and optimization.

Though there are over a dozen good books on such topics, most of these books are either too specialized with specific readership or too lengthy (often over 500 pages). This book fills in the gap with a compact and concise approach by focusing on the key concepts, algorithms, and techniques at an introductory level. The main approach of this book is informal, theorem-free, and practical. By using an informal approach all fundamental topics required for data mining and machine learning are covered, and the readers can gain such basic knowledge of all important algorithms with a focus on their key ideas, without worrying about any tedious, rigorous mathematical proofs. In addition, the practical approach provides about 30 worked examples in this book so that the readers can see how each step of the algorithms and techniques works. Thus, the readers can build their understanding and confidence gradually and in a step-by-step manner. Furthermore, with the minimal requirements of basic high school mathematics and some basic calculus, such an informal and practical style can also enable the readers to learn the contents by self-study and at their own pace.

This book is suitable for undergraduates and graduates to rapidly develop all the fundamental knowledge of data mining, machine learning, and optimization. It can also be used by students and researchers as a reference to review and refresh their knowledge in data mining, machine learning, optimization, computer science, and data science.

Xin-She Yang
January 2019 in London

Acknowledgments

I would like to thank all my students and colleagues who have given valuable feedback and comments on some of the contents and examples of this book. I also would like to thank my editors, J. Scott Bentley and Michael Lutz, and the staff at Elsevier for their professionalism. Last but not least, I thank my family for all the help and support.

Xin-She Yang
January 2019

Introduction to optimization

Contents

This book introduces the most fundamentals and algorithms related to optimization, data mining, and machine learning. The main requirement is some understanding of high-school mathematics and basic calculus; however, we will review and introduce some of the mathematical foundations in the first two chapters.

1.1 Algorithms

An algorithm is an iterative, step-by-step procedure for computation. The detailed procedure can be a simple description, an equation, or a series of descriptions in combination with equations. Finding the roots of a polynomial, checking if a natural number is a prime number, and generating random numbers are all algorithms.

1.1.1 Essence of an algorithm

In essence, an algorithm can be written as an iterative equation or a set of iterative equations. For example, to find a square root of $a > 0$, we can use the following iterative equation:

$$x_{k+1} = \frac{1}{2}\Big(x_k + \frac{a}{x_k}\Big), \tag{1.1}$$

where k is the iteration counter ($k = 0, 1, 2, \ldots$) starting with a random guess $x_0 = 1$.

Introduction to Algorithms for Data Mining and Machine Learning. https://doi.org/10.1016/B978-0-12-817216-2.00008-9

Example 1

As an example, if $x_0 = 1$ and $a = 4$, then we have

$$x_1 = \frac{1}{2}(1 + \frac{4}{1}) = 2.5. \tag{1.2}$$

Similarly, we have

$$x_2 = \frac{1}{2}(2.5 + \frac{4}{2.5}) = 2.05, \quad x_3 = \frac{1}{2}(2.05 + \frac{4}{2.05}) \approx 2.0061, \tag{1.3}$$

$$x_4 \approx 2.00000927, \tag{1.4}$$

which is very close to the true value of $\sqrt{4} = 2$. The accuracy of this iterative formula or algorithm is high because it achieves the accuracy of five decimal places after four iterations.

The convergence is very quick if we start from different initial values such as $x_0 = 10$ and even $x_0 = 100$. However, for an obvious reason, we cannot start with $x_0 = 0$ due to division by zero.

Find the root of $x = \sqrt{a}$ is equivalent to solving the equation

$$f(x) = x^2 - a = 0, \tag{1.5}$$

which is again equivalent to finding the roots of a polynomial $f(x)$. We know that Newton's root-finding algorithm can be written as

$$x_{k+1} = x_k - \frac{f(x_k)}{f'(x_k)}, \tag{1.6}$$

where $f'(x)$ is the first derivative or gradient of $f(x)$. In this case, we have $f'(x) = 2x$. Thus, Newton's formula becomes

$$x_{k+1} = x_k - \frac{(x_k^2 - a)}{2x_k}, \tag{1.7}$$

which can be written as

$$x_{k+1} = (x_k - \frac{x_k}{2}) + \frac{a}{2x_k} = \frac{1}{2}(x_k + \frac{a}{x_k}). \tag{1.8}$$

This is exactly what we have in Eq. (1.1).

Newton's method has rigorous mathematical foundations, which has a guaranteed convergence under certain conditions. However, in general, Eq. (1.6) is more general, and the gradient information $f'(x)$ is needed. In addition, for the formula to be valid, we must have $f'(x) \neq 0$.

1.1.2 Issues with algorithms

The advantage of the algorithm given in Eq. (1.1) is that it converges very quickly. However, careful readers may have asked: we know that $\sqrt{4} = \pm 2$, how can we find the other root -2 in addition to $+2$?

Even if we use different initial value $x_0 = 10$ or $x_0 = 0.5$, we can only reach $x_* = 2$, not -2.

What happens if we start with $x_0 < 0$? From $x_0 = -1$, we have

$$x_1 = \frac{1}{2}(-1 + \frac{4}{-1}) = -2.5, \quad x_2 = \frac{1}{2}(-2.5 + \frac{4}{-2.5}) = -2.05, \tag{1.9}$$

$$x_3 \approx -2.0061, \quad x_4 \approx -2.00000927, \tag{1.10}$$

which is approaching -2 very quickly. If we start from $x_0 = -10$ or $x_0 = -0.5$, then we can always get $x_* = -2$, not $+2$.

This highlights a key issue here: the final solution seems to depend on the initial starting point for this algorithm, which is true for many algorithms.

Now the relevant question is: how do we know where to start to get a particular solution? The general short answer is "we do not know". Thus, some knowledge of the problem under consideration or an educated guess may be useful to find the final solution.

In fact, most algorithms may depend on the initial configuration, and such algorithms are often carrying out search moves locally. Thus, this type of algorithm is often referred to as local search. A good algorithm should be able to "forget" its initial configuration though such algorithms may not exist at all for most types of problems.

What we need in general is the global search, which attempts to find final solutions that are less sensitive to the initial starting point(s).

Another important issue in our discussions is that the gradient information $f'(x)$ is necessary for some algorithms such as Newton's method given in Eq. (1.6). This poses certain requirements on the smoothness of the function $f(x)$. For example, we know that $|x|$ is not differentiable at $x = 0$. Thus, we cannot directly use Newton's method to find the roots of $f(x) = |x|x^2 - a = 0$ for $a > 0$. Some modifications are needed.

There are other issues related to algorithms such as the setting of parameters, the slow rate of convergence, condition numbers, and iteration structures. All these make algorithm designs and usage somehow challenging, and we will discuss these issues in more detail later in this book.

1.1.3 Types of algorithms

An algorithm can only do a specific computation task (at most a class of computational tasks), and no algorithms can do all the tasks. Thus, algorithms can be classified due to their purposes. An algorithm to find roots of a polynomial belongs to root-finding algorithms, whereas an algorithm for ranking a set of numbers belongs to sorting algorithms. There are many classes of algorithms for different purposes. Even for the same purpose such as sorting, there are many different algorithms such as the merge sort, bubble sort, quicksort, and others.

We can also categorize algorithms in terms of their characteristics. The root-finding algorithms we just introduced are deterministic algorithms because the final solutions are exactly the same if we start from the same initial guess. We obtain the same set of solutions every time we run the algorithm. On the other hand, we may introduce some randomization into the algorithm, for example, using purely random initial points. Every time we run the algorithm, we use a new random initial guess. In this case, the algorithm can have some nondeterministic nature, and such algorithms are referred to as stochastic. Sometimes, using randomness may be advantageous. For example, in the example of $\sqrt{4} = \pm 2$ using Eq. (1.1), random initial values (both positive and negative) can allow the algorithm to find both roots. In fact, a major trend in the modern metaheuristics is using some randomization to suit different purposes.

For algorithms to be introduced in this book, we are mainly concerned with algorithms for data mining, optimization, and machine learning. We use a relatively unified approach to link algorithms in data mining and machine learning to algorithms for optimization.

1.2 Optimization

Optimization is everywhere, from engineering design to business planning. After all, time and resources are limited, and optimal use of such valuable resources is crucial. In addition, designs of products have to maximize the performance, sustainability, and energy efficiency and to minimize the costs. Therefore, optimization is important for many applications.

1.2.1 A simple example

Let us start with a very simple example to design a container with volume capacity $V_0 = 10 \text{ m}^3$. As the main cost is related to the cost of materials, the main aim is to minimize the total surface area S.

The first thing we have to decide is the shape of the container (cylinder, cubic, sphere or ellipsoid, or more complex geometry). For simplicity, let us start with a cylindrical shape with radius r and height h (see Fig. 1.1).

The total surface area of a cylinder is

$$S = 2(\pi r^2) + 2\pi r h, \tag{1.11}$$

and the volume is

$$V = \pi r^2 h. \tag{1.12}$$

There are only two design variables r and h and one objective function S to be minimized. Obviously, if there is no capacity constraint, then we can choose not to build the container, and then the cost of materials is zero for $r = 0$ and $h = 0$. However,

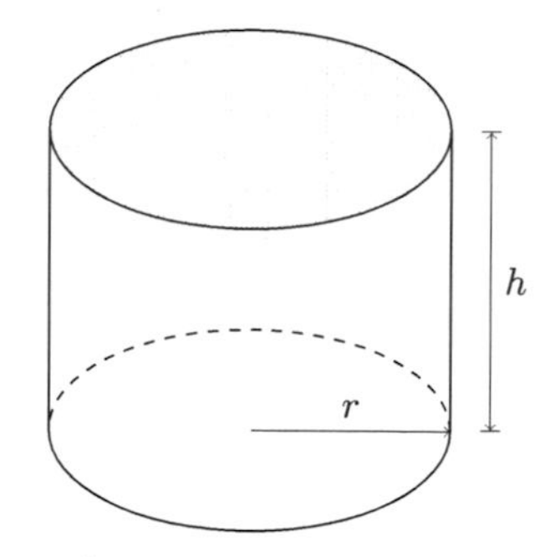

Figure 1.1 Design of a cylindric container.

the constraint requirement means that we have to build a container with fixed volume $V_0 = \pi r^2 h = 10 \text{ m}^3$. Therefore, this optimization problem can be written as

$$\text{minimize} \quad S = 2\pi r^2 + 2\pi r h, \tag{1.13}$$

subject to the equality constraint

$$\pi r^2 h = V_0 = 10. \tag{1.14}$$

To solve this problem, we can first try to use the equality constraint to reduce the number of design variables by solving h. So we have

$$h = \frac{V_0}{\pi r^2}. \tag{1.15}$$

Substituting it into (1.13), we get

$$\begin{aligned} S &= 2\pi r^2 + 2\pi r h \\ &= 2\pi r^2 + 2\pi r \frac{V_0}{\pi r^2} = 2\pi r^2 + \frac{2V_0}{r}. \end{aligned} \tag{1.16}$$

This is a univariate function. From basic calculus we know that the minimum or maximum can occur at the stationary point, where the first derivative is zero, that is,

$$\frac{\mathrm{d}S}{\mathrm{d}r} = 4\pi r - \frac{2V_0}{r^2} = 0, \tag{1.17}$$

which gives

$$r^3 = \frac{V_0}{2\pi}, \quad \text{or} \quad r = \sqrt[3]{\frac{V_0}{2\pi}}. \tag{1.18}$$

Thus, the height is

$$\frac{h}{r} = \frac{V_0/(\pi r^2)}{r} = \frac{V_0}{\pi r^3} = 2. \tag{1.19}$$

This means that the height is twice the radius: $h = 2r$. Thus, the minimum surface is

$$S_* = 2\pi r^2 + 2\pi r h = 2\pi r^2 + 2\pi r(2r) = 6\pi r^2$$
$$= 6\pi \Big(\frac{V_0}{2\pi}\Big)^{2/3} = \frac{6\pi}{\sqrt[3]{4\pi^2}} V_0^{2/3}. \tag{1.20}$$

For $V_0 = 10$, we have

$$r = \sqrt[3]{\frac{V_0}{(2\pi)}} = \sqrt[3]{\frac{10}{2\pi}} \approx 1.1675, \quad h = 2r = 2.335,$$

and the total surface area

$$S_* = 2\pi r^2 + 2\pi r h \approx 25.69.$$

It is worth pointing out that this optimal solution is based on the assumption or requirement to design a cylindrical container. If we decide to use a sphere with radius R, we know that its volume and surface area is

$$V_0 = \frac{4\pi}{3} R^3, \quad S = 4\pi R^2. \tag{1.21}$$

We can solve R directly

$$R^3 = \frac{3V_0}{4\pi}, \quad \text{or } R = \sqrt[3]{\frac{3V_0}{4\pi}}, \tag{1.22}$$

which gives the surface area

$$S = 4\pi \Big(\frac{3V_0}{4\pi}\Big)^{2/3} = \frac{4\pi \sqrt[3]{9}}{\sqrt[3]{16\pi^2}} V_0^{2/3}. \tag{1.23}$$

Since $6\pi/\sqrt[3]{4\pi^2} \approx 5.5358$ and $4\pi \sqrt[3]{9}/\sqrt[3]{16\pi^2} \approx 4.83598$, we have $S < S_*$, that is, the surface area of a sphere is smaller than the minimum surface area of a cylinder with the same volume. In fact, for the same $V_0 = 10$, we have

$$S(\text{sphere}) = \frac{4\pi \sqrt[3]{9}}{\sqrt[3]{16\pi^2}} V_0^{2/3} \approx 22.47, \tag{1.24}$$

which is smaller than $S_* = 25.69$ for a cylinder.

This highlights the importance of the choice of design type (here in terms of shape) before we can do any truly useful optimization. Obviously, there are many other factors that can influence the choice of design, including the manufacturability of the design, stability of the structure, ease of installation, space availability, and so on. For a container, in most applications, a cylinder may be much easier to produce than a sphere, and thus the overall cost may be lower in practice. Though there are so many factors to be considered in engineering design, for the purpose of optimization, here we will only focus on the improvement and optimization of a design with well-posed mathematical formulations.

1.2.2 General formulation of optimization

Whatever the real-world applications may be, it is usually possible to formulate an optimization problem in a generic form [49,53,160]. All optimization problems with explicit objectives can in general be expressed as a nonlinearly constrained optimization problem

$$\begin{aligned} &\text{maximize/minimize} \quad f(\boldsymbol{x}), \ \boldsymbol{x} = (x_1, x_2, \ldots, x_D)^T \in \mathbb{R}^D, \\ &\text{subject to } \phi_j(\boldsymbol{x}) = 0 \ \ (j = 1, 2, \ldots, M), \\ &\qquad\qquad \psi_k(\boldsymbol{x}) \le 0 \ \ (k = 1, \ldots, N), \end{aligned} \tag{1.25}$$

where $f(\boldsymbol{x})$, $\phi_j(\boldsymbol{x})$, and $\psi_k(\boldsymbol{x})$ are scalar functions of the design vector $\boldsymbol{x}$. Here the components x_i of $\boldsymbol{x} = (x_1, \ldots, x_D)^T$ are called design or decision variables, and they can be either continuous, discrete, or a mixture of these two. The vector $\boldsymbol{x}$ is often called the decision vector, which varies in a D-dimensional space $\mathbb{R}^D$.

It is worth pointing out that we use a column vector here for $\boldsymbol{x}$ (thus with transpose T). We can also use a row vector $\boldsymbol{x} = (x_1, \ldots, x_D)$ and the results will be the same. Different textbooks may use slightly different formulations. Once we are aware of such minor variations, it should cause no difficulty or confusion.

In addition, the function $f(\boldsymbol{x})$ is called the objective function or cost function, $\phi_j(\boldsymbol{x})$ are constraints in terms of M equalities, and $\psi_k(\boldsymbol{x})$ are constraints written as N inequalities. So there are $M + N$ constraints in total. The optimization problem formulated here is a nonlinear constrained problem. Here the inequalities $\psi_k(\boldsymbol{x}) \le 0$ are written as "less than", and they can also be written as "greater than" via a simple transformation by multiplying both sides by -1.

The space spanned by the decision variables is called the search space $\mathbb{R}^D$, whereas the space formed by the values of the objective function is called the objective or response space, and sometimes the landscape. The optimization problem essentially maps the domain $\mathbb{R}^D$ or the space of decision variables into the solution space $\mathbb{R}$ (or the real axis in general).

The objective function $f(\boldsymbol{x})$ can be either linear or nonlinear. If the constraints ϕ_j and ψ_k are all linear, it becomes a linearly constrained problem. Furthermore, when ϕ_j, ψ_k, and the objective function $f(\boldsymbol{x})$ are all linear, then it becomes a linear programming problem [35]. If the objective is at most quadratic with linear constraints, then it is called a quadratic programming problem. If all the values of the decision variables can be only integers, then this type of linear programming is called integer programming or integer linear programming.

On the other hand, if no constraints are specified and thus x_i can take any values in the real axis (or any integers), then the optimization problem is referred to as an unconstrained optimization problem.

As a very simple example of optimization problems without any constraints, we discuss the search of the maxima or minima of a univariate function.

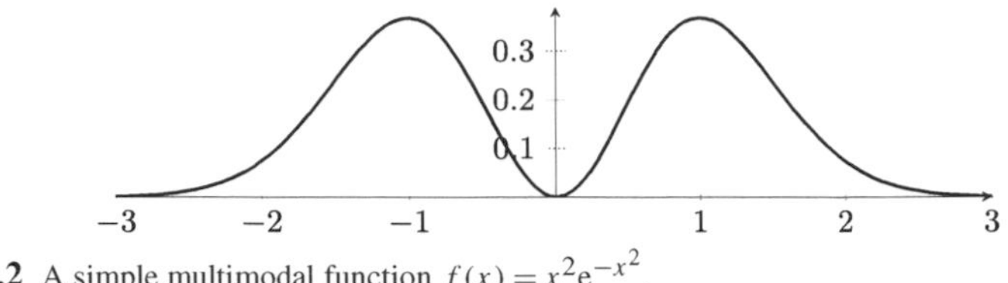

Figure 1.2 A simple multimodal function $f(x) = x^2 e^{-x^2}$.

Example 2

For example, to find the maximum of a univariate function $f(x)$

$$f(x) = x^2 e^{-x^2}, \qquad -\infty < x < \infty, \tag{1.26}$$

is a simple unconstrained problem, whereas the following problem is a simple constrained minimization problem:

$$f(x_1, x_2) = x_1^2 + x_1 x_2 + x_2^2, \qquad (x_1, x_2) \in \mathbb{R}^2, \tag{1.27}$$

subject to

$$x_1 \geq 1, \qquad x_2 - 2 = 0. \tag{1.28}$$

It is worth pointing out that the objectives are explicitly known in all the optimization problems to be discussed in this book. However, in reality, it is often difficult to quantify what we want to achieve, but we still try to optimize certain things such as the degree of enjoyment or service quality on holiday. In other cases, it may be impossible to write the objective function in any explicit form mathematically.

From basic calculus we know that, for a given curve described by $f(x)$, its gradient $f'(x)$ describes the rate of change. When $f'(x) = 0$, the curve has a horizontal tangent at that particular point. This means that it becomes a point of special interest. In fact, the maximum or minimum of a curve occurs at

$$f'(x_*) = 0, \tag{1.29}$$

which is a critical condition or stationary condition. The solution x_* to this equation corresponds to a stationary point, and there may be multiple stationary points for a given curve.

To see if it is a maximum or minimum at $x = x_*$, we have to use the information of its second derivative $f''(x)$. In fact, $f''(x_*) > 0$ corresponds to a minimum, whereas $f''(x_*) < 0$ corresponds to a maximum. Let us see a concrete example.

Example 3

To find the minimum of $f(x) = x^2 e^{-x^2}$ (see Fig. 1.2), we have the stationary condition $f'(x) = 0$ or

$$f'(x) = 2x \times e^{-x^2} + x^2 \times (-2x) e^{-x^2} = 2(x - x^3) e^{-x^2} = 0.$$

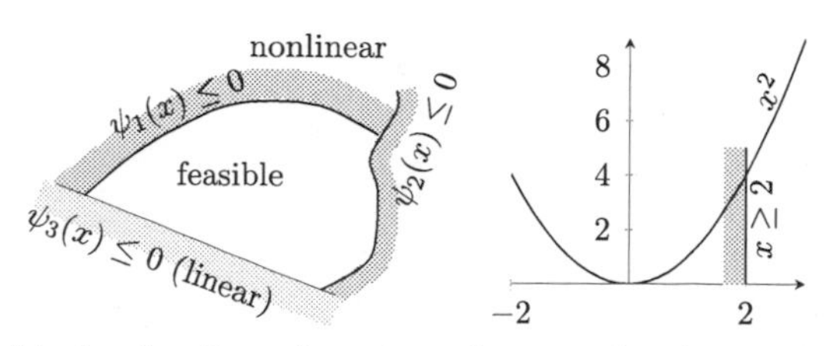

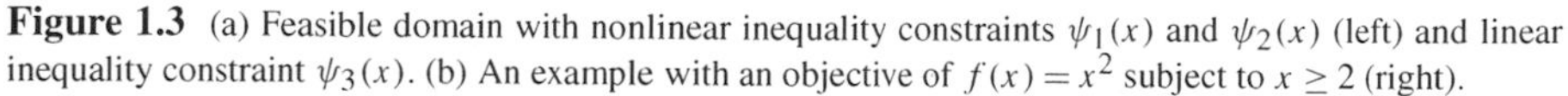
Figure 1.3 (a) Feasible domain with nonlinear inequality constraints $\psi_1(x)$ and $\psi_2(x)$ (left) and linear inequality constraint $\psi_3(x)$. (b) An example with an objective of $f(x) = x^2$ subject to $x \geq 2$ (right).

As $e^{-x^2} > 0$, we have

$$x(1 - x^2) = 0, \quad \text{or} \quad x = 0 \text{ and } x = \pm 1.$$

The second derivative is given by

$$f''(x) = 2e^{-x^2}(1 - 5x^2 + 2x^4),$$

which is an even function with respect to x.

So at $x = \pm 1$, $f''(\pm 1) = 2[1 - 5(\pm 1)^2 + 2(\pm 1)^4]e^{-(\pm 1)^2} = -4e^{-1} < 0$. Thus, there are two maxima that occur at $x_* = \pm 1$ with $f_{max} = e^{-1}$. At $x = 0$, we have $f''(0) = 2 > 0$, thus the minimum of $f(x)$ occurs at $x_* = 0$ with $f_{min}(0) = 0$.

Whatever the objective is, we have to evaluate it many times. In most cases, the evaluations of the objective functions consume a substantial amount of computational power (which costs money) and design time. Any efficient algorithm that can reduce the number of objective evaluations saves both time and money.

In mathematical programming, there are many important concepts, and we will first introduce a few related concepts: feasible solutions, optimality criteria, the strong local optimum, and weak local optimum.

1.2.3 Feasible solution

A point $\boldsymbol{x}$ that satisfies all the constraints is called a feasible point and thus is a feasible solution to the problem. The set of all feasible points is called the feasible region (see Fig. 1.3).

For example, we know that the domain $f(x) = x^2$ consists of all real numbers. If we want to minimize $f(x)$ without any constraint, all solutions such as $x = -1$, $x = 1$, and $x = 0$ are feasible. In fact, the feasible region is the whole real axis. Obviously, $x = 0$ corresponds to $f(0) = 0$ as the true minimum.

However, if we want to find the minimum of $f(x) = x^2$ subject to $x \geq 2$, then it becomes a constrained optimization problem. The points such as $x = 1$ and $x = 0$ are no longer feasible because they do not satisfy $x \geq 2$. In this case the feasible solutions are all the points that satisfy $x \geq 2$. So $x = 2$, $x = 100$, and $x = 10^8$ are all feasible. It is obvious that the minimum occurs at $x = 2$ with $f(2) = 2^2 = 4$, that is, the optimal solution for this problem occurs at the boundary point $x = 2$ (see Fig. 1.3).

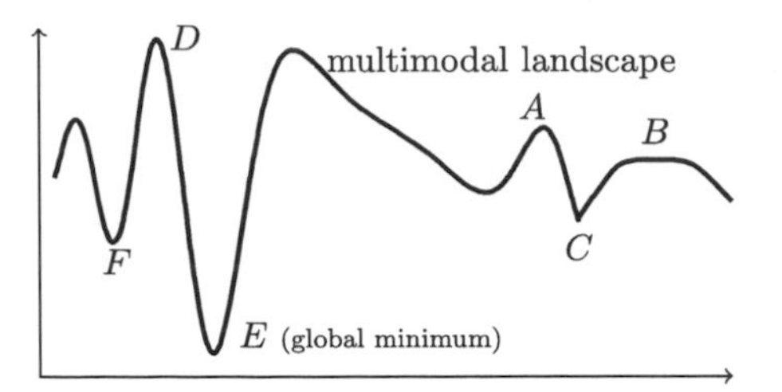

Figure 1.4 Local optima, weak optima, and global optimality.

1.2.4 Optimality criteria

A point $\boldsymbol{x}_*$ is called a strong local maximum of the nonlinearly constrained optimization problem if $f(\boldsymbol{x})$ is defined in a δ-neighborhood $N(\boldsymbol{x}_*, \delta)$ and satisfies $f(\boldsymbol{x}_*) > f(\boldsymbol{u})$ for $\boldsymbol{u} \in N(\boldsymbol{x}_*, \delta)$, where $\delta > 0$ and $\boldsymbol{u} \neq \boldsymbol{x}_*$. If $\boldsymbol{x}_*$ is not a strong local maximum, then the inclusion of equality in the condition $f(\boldsymbol{x}_*) \geq f(\boldsymbol{u})$ for all $\boldsymbol{u} \in N(\boldsymbol{x}_*, \delta)$ defines the point $\boldsymbol{x}_*$ as a weak local maximum (see Fig. 1.4). The local minima can be defined in a similar manner when $>$ and $\geq$ are replaced by $<$ and $\leq$, respectively.

Fig. 1.4 shows various local maxima and minima. Point A is a strong local maximum, whereas point B is a weak local maximum because there are many (in fact, infinite) different values of $\boldsymbol{x}$ that will lead to the same value of $f(\boldsymbol{x}_*)$. Point D is the global maximum, and point E is the global minimum. In addition, point F is a strong local minimum. However, point C is a strong local minimum, but it has a discontinuity in $f'(\boldsymbol{x}_*)$. So the stationary condition for this point $f'(\boldsymbol{x}_*) = 0$ is not valid. We will not deal with these types of minima or maxima in detail.

As we briefly mentioned before, for a smooth curve $f(x)$, optimal solutions usually occur at stationary points where $f'(x) = 0$. This is not always the case because optimal solutions can also occur at the boundary, as we have seen in the previous example of minimizing $f(x) = x^2$ subject to $x \geq 2$. In our present discussion, we will assume that both $f(\boldsymbol{x})$ and $f'(\boldsymbol{x})$ are always continuous or $f(\boldsymbol{x})$ is everywhere twice continuously differentiable. Obviously, the information of $f'(x)$ is not sufficient to determine whether a stationary point is a local maximum or minimum. Thus, higher-order derivatives such as $f''(x)$ are needed, but we do not make any assumption at this stage. We will further discuss this in detail in the next section.

1.3 Unconstrained optimization

Optimization problems can be classified as either unconstrained or constrained. Unconstrained optimization problems can in turn be subdivided into univariate and multivariate problems.

1.3.1 Univariate functions

The simplest optimization problem without any constraints is probably the search for the maxima or minima of a univariate function $f(x)$. For unconstrained optimization problems, the optimality occurs at the critical points given by the stationary condition $f'(x) = 0$.

However, this stationary condition is just a necessary condition, but it is not a sufficient condition. If $f'(x_*) = 0$ and $f''(x_*) > 0$, it is a local minimum. Conversely, if $f'(x_*) = 0$ and $f''(x_*) < 0$, then it is a local maximum. However, if $f'(x_*) = 0$ and $f''(x_*) = 0$, care should be taken because $f''(x)$ may be indefinite (both positive and negative) when $x \to x_*$, then x_* corresponds to a saddle point.

For example, for $f(x) = x^3$, we have

$$f'(x) = 3x^2, \quad f''(x) = 6x. \tag{1.30}$$

The stationary condition $f'(x) = 3x^2 = 0$ gives $x_* = 0$. However, we also have

$$f''(x_*) = f''(0) = 0.$$

In fact, $f(x) = x^3$ has a saddle point $x_* = 0$ because $f'(0) = 0$ but f'' changes sign from $f''(0+) > 0$ to $f''(0-) < 0$ as x moves from positive to negative.

Example 4

For example, to find the maximum or minimum of a univariate function

$$f(x) = 3x^4 - 4x^3 - 12x^2 + 9, \qquad -\infty < x < \infty,$$

we first have to find its stationary points x_* when the first derivative $f'(x)$ is zero, that is,

$$f'(x) = 12x^3 - 12x^2 - 24x = 12(x^3 - x^2 - 2x) = 0.$$

Since $f'(x) = 12(x^3 - x^2 - 2x) = 12x(x+1)(x-2) = 0$, we have

$$x_* = -1, \quad x_* = 2, \quad x_* = 0.$$

The second derivative of $f(x)$ is simply

$$f''(x) = 36x^2 - 24x - 24.$$

From the basic calculus we know that the maximum requires $f''(x_*) \leq 0$ whereas the minimum requires $f''(x_*) \geq 0$.

At $x_* = -1$, we have

$$f''(-1) = 36(-1)^2 - 24(-1) - 24 = 36 > 0,$$

so this point corresponds to a local minimum

$$f(-1) = 3(-1)^4 - 4(-1)^3 - 12(-1)^2 + 9 = 4.$$

Similarly, at $x_* = 2$, $f''(x_*) = 72 > 0$, and thus we have another local minimum

$$f(x_*) = -23.$$

However, at $x_* = 0$, we have $f'(0) = -24 < 0$, which corresponds to a local maximum $f(0) = 9$. However, this maximum is not a global maximum because the global maxima for $f(x)$ occur at $x = \pm\infty$.

The global minimum occurs at $x_* = 2$ with $f(2) = -23$.

The maximization of a function $f(x)$ can be converted into the minimization of $A - f(x)$, where A is usually a large positive number (though $A = 0$ will do). For example, we know the maximum of $f(x) = e^{-x^2}$, $x \in (-\infty, \infty)$, is 1 at $x_* = 0$. This problem can be converted to the minimization of $-f(x)$. For this reason, the optimization problems can be expressed as either minimization or maximization depending on the context and convenience of formulations.

In fact, in the optimization literature, some books formulate all the optimization problems in terms of maximization, whereas others write these problems in terms of minimization, though they are in essence dealing with the same problems.

1.3.2 Multivariate functions

We can extend the optimization procedure for univariate functions to multivariate functions using partial derivatives and relevant conditions. Let us start with the example

$$\text{minimize} \quad f(x, y) = x^2 + y^2, \quad x, y \in \mathbb{R}. \tag{1.31}$$

It is obvious that $x = 0$ and $y = 0$ is a minimum solution because $f(0, 0) = 0$. The question is how to solve this problem formally. We can extend the stationary condition to partial derivatives, and we have $\frac{\partial f}{\partial x} = 0$ and $\frac{\partial f}{\partial y} = 0$. In this case, we have

$$\frac{\partial f}{\partial x} = 2x + 0 = 0, \quad \frac{\partial f}{\partial y} = 0 + 2y = 0. \tag{1.32}$$

The solution is obviously $x_* = 0$ and $y_* = 0$.

Now how do we know that it corresponds to a maximum or minimum? If we try to use the second derivatives, we have four different partial derivatives such as f_{xx} and f_{yy}, and which one should we use? In fact, we need to define the Hessian matrix from these second partial derivatives, and we have

$$\boldsymbol{H} = \begin{pmatrix} f_{xx} & f_{xy} \\ f_{yx} & f_{yy} \end{pmatrix} = \begin{pmatrix} \frac{\partial^2 f}{\partial x^2} & \frac{\partial^2 f}{\partial x \partial y} \\ \frac{\partial^2 f}{\partial y \partial x} & \frac{\partial^2 f}{\partial y^2} \end{pmatrix}. \tag{1.33}$$

Since

$$\frac{\partial^2 f}{\partial x \partial y} = \frac{\partial^2 f}{\partial y \partial x}, \tag{1.34}$$

we can conclude that the Hessian matrix is always symmetric. In the case of $f(x, y) = x^2 + y^2$, it is easy to check that the Hessian matrix is

$$\boldsymbol{H} = \begin{pmatrix} 2 & 0 \\ 0 & 2 \end{pmatrix}. \tag{1.35}$$

Mathematically speaking, if $\boldsymbol{H}$ is positive definite, then the stationary point (x_*, y_*) corresponds to a local minimum. Similarly, if $\boldsymbol{H}$ is negative definite, then the stationary point corresponds to a maximum. The definiteness of a symmetric matrix is controlled by its eigenvalues. For this simple diagonal matrix $\boldsymbol{H}$, its eigenvalues are its two diagonal entries 2 and 2. As both eigenvalues are positive, this matrix is positive definite. Since the Hessian matrix here does not involve any x or y, it is always positive definite in the whole search domain $(x, y) \in \mathbb{R}^2$, so we can conclude that the solution at point $(0, 0)$ is the global minimum.

Obviously, this is a particular case. In general, the Hessian matrix depends on the independent variables, but the definiteness test conditions still apply. That is, positive definiteness of a stationary point means a local minimum. Alternatively, for bivariate functions, we can define the determinant of the Hessian matrix in Eq. (1.33) as

$$\Delta = \det(\boldsymbol{H}) = f_{xx} f_{yy} - (f_{xy})^2. \tag{1.36}$$

At the stationary point (x_*, y_*), if $\Delta > 0$ and $f_{xx} > 0$, then (x_*, y_*) is a local minimum. If $\Delta > 0$ but $f_{xx} < 0$, then it is a local maximum. If $\Delta = 0$, then it is inconclusive, and we have to use other information such as higher-order derivatives. However, if $\Delta < 0$, then it is a saddle point. A saddle point is a special point where a local minimum occurs along one direction, whereas the maximum occurs along another (orthogonal) direction.

Example 5

To minimize $f(x, y) = (x - 1)^2 + x^2 y^2$, we have

$$\frac{\partial f}{\partial x} = 2(x - 1) + 2xy^2 = 0, \quad \frac{\partial f}{\partial y} = 0 + 2x^2 y = 0. \tag{1.37}$$

The second condition gives $y = 0$ or $x = 0$. Substituting $y = 0$ into the first condition, we have $x = 1$. However, $x = 0$ does not satisfy the first condition. Therefore, we have a solution $x_* = 1$ and $y_* = 0$.

For our example with $f = (x - 1)^2 + x^2 y^2$, we have

$$\frac{\partial^2 f}{\partial x^2} = 2y^2 + 2, \quad \frac{\partial^2 f}{\partial x \partial y} = 4xy, \quad \frac{\partial^2 f}{\partial y \partial x} = 4xy, \quad \frac{\partial^2 f}{\partial y^2} = 2x^2, \tag{1.38}$$

and thus we have

$$\boldsymbol{H} = \begin{pmatrix} 2y^2+2 & 4xy \\ 4xy & 2x^2 \end{pmatrix}. \tag{1.39}$$

At the stationary point $(x_*, y_*) = (1, 0)$, the Hessian matrix becomes

$$\boldsymbol{H} = \begin{pmatrix} 2 & 0 \\ 0 & 2 \end{pmatrix},$$

which is positive definite because its double eigenvalues 2 are positive. Alternatively, we have $\Delta = 4 > 0$ and $f_{xx} = 2 > 0$. Therefore, $(1, 0)$ is a local minimum.

In fact, for a multivariate function $f(x_1, x_2, \ldots, x_n)$ in an n-dimensional space, the stationary condition can be extended to

$$\boldsymbol{G} = \nabla f = (\frac{\partial f}{\partial x_1}, \frac{\partial f}{\partial x_2}, \ldots, \frac{\partial f}{\partial x_n})^T = 0, \tag{1.40}$$

where $\boldsymbol{G}$ is called the gradient vector. The second derivative test becomes the definiteness of the Hessian matrix

$$\boldsymbol{H} = \begin{pmatrix} \frac{\partial^2 f}{\partial {x_1}^2} & \frac{\partial^2 f}{\partial x_1 \partial x_2} & \cdots & \frac{\partial^2 f}{\partial x_1 \partial x_n} \\ \frac{\partial^2 f}{\partial x_2 \partial x_1} & \frac{\partial^2 f}{\partial {x_2}^2} & \cdots & \frac{\partial^2 f}{\partial x_2 \partial x_n} \\ \vdots & \vdots & \ddots & \vdots \\ \frac{\partial^2 f}{\partial x_n \partial x_1} & \frac{\partial^2 f}{\partial x_n \partial x_2} & \cdots & \frac{\partial^2 f}{\partial {x_n}^2} \end{pmatrix}. \tag{1.41}$$

At the stationary point defined by $\boldsymbol{G} = \nabla f = 0$, the positive definiteness of $\boldsymbol{H}$ gives a local minimum, whereas the negative definiteness corresponds to a local maximum. In essence, the eigenvalues of the Hessian matrix $\boldsymbol{H}$ determine the local behavior of the function. As we mentioned before, if $\boldsymbol{H}$ is positive semidefinite, then it corresponds to a local minimum.

1.4 Nonlinear constrained optimization

As most real-world problems are nonlinear, nonlinear mathematical programming forms an important part of mathematical optimization methods. A broad class of nonlinear programming problems is about the minimization or maximization of $f(\boldsymbol{x})$ subject to no constraints, and another important class is the minimization of a quadratic objective function subject to nonlinear constraints. There are many other nonlinear programming problems as well.

Nonlinear programming problems are often classified according to the convexity of the defining functions. An interesting property of a convex function f is that the vanishing of the gradient $\nabla f(\boldsymbol{x}_*) = 0$ guarantees that the point x_* is a global minimum or maximum of f. We will introduce the concept of convexity in the next chapter. If a function is not convex or concave, then it is much more difficult to find its global minima or maxima.

1.4.1 Penalty method

For the simple function optimization with equality and inequality constraints, a common method is the penalty method. For the optimization problem

$$\text{minimize} \quad f(\boldsymbol{x}), \quad \boldsymbol{x} = (x_1, \ldots, x_n)^T \in \mathbb{R}^n,$$

$$\text{subject to } \phi_i(\boldsymbol{x}) = 0, \ (i = 1, \ldots, M), \quad \psi_j(\boldsymbol{x}) \le 0, \ (j = 1, \ldots, N), \tag{1.42}$$

the idea is to define a penalty function so that the constrained problem is transformed into an unconstrained problem. Now we define

$$\Pi(\boldsymbol{x}, \mu_i, \nu_j) = f(\boldsymbol{x}) + \sum_{i=1}^{M} \mu_i \phi_i^2(\boldsymbol{x}) + \sum_{j=1}^{N} \nu_j \max\{0, \psi_j(\boldsymbol{x})\}^2, \tag{1.43}$$

where $\mu_i \gg 1$ and $\nu_j \ge 0$.

For example, let us solve the following minimization problem:

$$\text{minimize} \quad f(x) = 40(x-1)^2, \quad x \in \mathbb{R}, \quad \text{subject to} \quad g(x) = x - a \ge 0, \tag{1.44}$$

where a is a given value. Obviously, without this constraint, the minimum value occurs at $x = 1$ with $f_{\min} = 0$. If $a < 1$, then the constraint will not affect the result. However, if $a > 1$, then the minimum should occur at the boundary $x = a$ (which can be obtained by inspecting or visualizing the objective function and the constraint). Now we can define a penalty function $\Pi(x)$ using a penalty parameter $\mu \gg 1$. We have

$$\Pi(x, \mu) = f(x) + \mu[g(x)]^2 = 40(x-1)^2 + \mu(x-a)^2, \tag{1.45}$$

which converts the original constrained optimization problem into an unconstrained problem. From the stationarity condition $\Pi'(x) = 0$ we have

$$80(x-1) - 2\mu(x-a) = 0, \qquad \text{or} \quad x_* = \frac{40 - \mu a}{40 - \mu}. \tag{1.46}$$

For a particular case $a = 1$, we have $x_* = 1$, and the result does not depend on μ. However, in the case of $a > 1$ (say, $a = 5$), the result will depend on μ. When $a = 5$ and $\mu = 100$, we have $x_* = 40 - 100 \times 5/40 - 100 = 7.6667$. If $\mu = 1000$, then this gives $50 - 1000 * 5/40 - 1000 = 5.1667$. Both values are far from the exact solution $x_{\text{true}} = a = 5$. If we use $\mu = 10^4$, then we have $x_* \approx 5.0167$. Similarly, for $\mu = 10^5$, we have $x_* \approx 5.00167$. This clearly demonstrates that the solution in general depends on μ. However, it is very difficult to use extremely large values without causing extra computational difficulties.

Ideally, the formulation using the penalty method should be properly designed so that the results will not depend on the penalty coefficient, or at least the dependence should be sufficiently weak.

1.4.2 Lagrange multipliers

Another powerful method without the limitation of using large μ is the method of Lagrange multipliers. Suppose we want to minimize a function $f(\boldsymbol{x})$:

$$\text{minimize } f(\boldsymbol{x}), \qquad \boldsymbol{x} = (x_1, \ldots, x_n)^T \in \mathbb{R}^n, \tag{1.47}$$

subject to the nonlinear equality constraint

$$h(\boldsymbol{x}) = 0. \tag{1.48}$$

Then we can combine the objective function $f(\boldsymbol{x})$ with the equality to form the new function, called the Lagrangian,

$$\Pi = f(\boldsymbol{x}) + \lambda h(\boldsymbol{x}), \tag{1.49}$$

where λ is the Lagrange multiplier, which is an unknown scalar to be determined.

This again converts the constrained optimization into an unconstrained problem for $\Pi(\boldsymbol{x})$, which is the beauty of this method. If we have M equalities

$$h_j(\boldsymbol{x}) = 0 \qquad (j = 1, \ldots, M), \tag{1.50}$$

then we need M Lagrange multipliers λ_j $(j = 1, \ldots, M)$. We thus have

$$\Pi(x, \lambda_j) = f(\boldsymbol{x}) + \sum_{j=1}^{M} \lambda_j h_j(\boldsymbol{x}). \tag{1.51}$$

The requirement of stationary conditions leads to

$$\frac{\partial \Pi}{\partial x_i} = \frac{\partial f}{\partial x_i} + \sum_{j=1}^{M} \lambda_j \frac{\partial h_j}{\partial x_i} \; (i = 1, \ldots, n), \qquad \frac{\partial \Pi}{\partial \lambda_j} = h_j = 0 \; (j = 1, \ldots, M). \tag{1.52}$$

These $M + n$ equations determine the n-component $\boldsymbol{x}$ and M Lagrange multipliers. As $\frac{\partial \Pi}{\partial g_j} = \lambda_j$, we can consider λ_j as the rate of the change of Π as a functional of h_j.

Example 6

For the well-known monkey surface $f(x, y) = x^3 - 3xy^2$, the function does not have a unique maximum or minimum. In fact, the point $x = y = 0$ is a saddle point. However, if we impose an extra equality $x - y^2 = 1$, we can formulate an optimization problem as

$$\text{minimize } f(x, y) = x^3 - 3xy^2, \quad (x, y) \in \mathbb{R}^2,$$

subject to

$$h(x, y) = x - y^2 = 1.$$

Now we can define

$$\Phi = f(x, y) + \lambda h(x, y) = x^3 - 3xy^2 + \lambda(x - y^2 - 1).$$

The stationary conditions become

$$\frac{\partial \Phi}{\partial x} = 3x^2 - 3y^2 + \lambda = 0, \qquad \frac{\partial \Phi}{\partial y} = 0 - 6xy + (-2\lambda y) = 0,$$

$$\frac{\partial \Phi}{\partial \lambda} = x - y^2 - 1 = 0.$$

The second condition $-6xy - 2\lambda y = -2y(3x + \lambda) = 0$ implies that $y = 0$ or $\lambda = -3x$.

- If $y = 0$, then the third condition $x - y^2 - 1 = 0$ gives $x = 1$. The first condition $3x^2 + 3y^2 - \lambda = 0$ leads to $\lambda = -3$. Therefore, $x = 1$ and $y = 0$ is an optimal solution with $f_{\min} = 1$. It is straightforward to verify that this solution corresponds to a minimum (not a maximum).
- If $\lambda = -3x$, then the first condition becomes $3x^2 - 3y^2 - 3x = 0$. Substituting $x = y^2 + 1$ (from the third condition), we have

$$3(y^2 + 1)^2 - 3y^2 - 3(y^2 + 1) = 0, \qquad \text{or} \qquad 3(y^4 + 2) = 0.$$

This equation has no solution in the real domain. Therefore, the optimality occurs at $(1, 0)$ with $f_{\min} = 1$.

1.4.3 Karush–Kuhn–Tucker conditions

There is a counterpart of the Lagrange multipliers for nonlinear optimization with inequality constraints. The Karush–Kuhn–Tucker (KKT) conditions concern the requirement for a solution to be optimal in nonlinear programming [111].

Let us know focus on the nonlinear optimization problem

$$\begin{aligned} &\text{minimize } f(\boldsymbol{x}), \ \boldsymbol{x} \in \mathbb{R}^n, \\ &\text{subject to } \phi_i(\boldsymbol{x}) = 0 \ (i = 1, \ldots, M), \quad \psi_j(\boldsymbol{x}) \leq 0 \ (j = 1, \ldots, N). \end{aligned} \tag{1.53}$$

If all the functions are continuously differentiable at a local minimum $\boldsymbol{x}_*$, then there exist constants $\lambda_0, \lambda_1, \ldots, \lambda_q$ and $\mu_1, \ldots, \mu_p$ such that

$$\lambda_0 \nabla f(\boldsymbol{x}_*) + \sum_{i=1}^{M} \mu_i \nabla \phi_i(\boldsymbol{x}_*) + \sum_{j=1}^{N} \lambda_j \nabla \psi_j(\boldsymbol{x}_*) = 0, \tag{1.54}$$

$$\psi_j(\boldsymbol{x}_*) \leq 0, \quad \lambda_j \psi_j(\boldsymbol{x}_*) = 0 \ (j = 1, 2, \ldots, N), \tag{1.55}$$

where $\lambda_j \geq 0$ $(i = 0, 1, \ldots, N)$. The constants satisfy $\sum_{j=0}^{N} \lambda_j + \sum_{i=1}^{M} |\mu_i| \geq 0$. This is essentially a generalized method of the Lagrange multipliers. However, there is a possibility of degeneracy when $\lambda_0 = 0$ under certain conditions.

It is worth pointing out that such KKT conditions can be useful to prove theorems and sometimes useful to gain insight into certain types of problems. However, they are not really helpful in practice in the sense that they do not give any indication where the optimal solutions may lie in the search domain so as to guide the search process.

Optimization problems, especially highly nonlinear multimodal problems, are usually difficult to solve. However, if we are mainly concerned about local optimal or suboptimal solutions (not necessarily about global optimal solutions), there are relatively efficient methods such as interior-point methods, trust-region methods, the simplex method, sequential quadratic programming, and swarm intelligence-based methods [151]. All these methods have been implemented in a diverse range of software packages. Interested readers can refer to more advanced literature.

1.5 Notes on software

Though there many different algorithms for optimization, most software packages and programming languages have some sort of optimization capabilities due to the popularity and relevance of optimization in many applications. For example, Wikipedia has some extensive lists of

- optimization software,[1]
- data mining and machine learning,[2]
- deep learning software.[3]

There is a huge list of software packages and internet resources; it requires a lengthy book to cover most of it, which is not our intention here. Interested readers can refer to them for more detail.

[1] https://en.wikipedia.org/wiki/List_of_optimization_software.

[2] https://en.wikipedia.org/wiki/Category:Data_mining_and_machine_learning_software.

[3] https://en.wikipedia.org/wiki/Comparison_of_deep_learning_software.

Mathematical foundations

Contents

Though the main requirement of this book is basic calculus, we will still briefly review some basic concepts concerning functions and basic calculus and then introduce some new concepts. The readers can skip this chapter if they are already familiar with such topics.

Introduction to Algorithms for Data Mining and Machine Learning. https://doi.org/10.1016/B978-0-12-817216-2.00009-0

2.1 Convexity

2.1.1 Linear and affine functions

Generally speaking, a function is a mapping from independent variables or inputs to a dependent variable or variables/outputs. For example, the function

$$f(x, y) = x^2 + y^2 + xy, \tag{2.1}$$

depends on two independent variables. This function maps the domain $\mathbb{R}^2$ (for $-\infty < x < \infty$ and $-\infty < y < \infty$) to f on the real axis as its range. So we use the notation $f : \mathbb{R}^2 \to \mathbb{R}$ to denote this.

In general, a function $f(x, y, z, \ldots)$ maps n independent variables to m dependent variables, and we use the notation $f : \mathbb{R}^n \to \mathbb{R}^m$ to mean that the domain of the function is a subset of $\mathbb{R}^n$, whereas its range is a subset of $\mathbb{R}^m$. The domain of a function is sometimes denoted by dom(f) or dom f.

The inputs or independent variables can often be written as a vector. For simplicity, we often use a vector $\boldsymbol{x} = (x, y, z, \ldots)^T = (x_1, x_2, \ldots, x_n)^T$ for multiple variables. Therefore, $f(\boldsymbol{x})$ is often used to mean $f(x, y, z, \ldots)$ or $f(x_1, x_2, \ldots, x_n)$.

A function $\mathcal{L}(\boldsymbol{x})$ is called linear if

$$\mathcal{L}(\boldsymbol{x} + \boldsymbol{y}) = \mathcal{L}(\boldsymbol{x}) + \mathcal{L}(\boldsymbol{y}) \quad \text{and} \quad \mathcal{L}(\alpha \boldsymbol{x}) = \alpha \mathcal{L}(\boldsymbol{x}) \tag{2.2}$$

for any vectors $\boldsymbol{x}$ and $\boldsymbol{y}$ and any scalar $\alpha \in \mathbb{R}$.

Example 7

To see if $f(\boldsymbol{x}) = f(x_1, x_2) = 2x_1 + 3x_2$ is linear, we use

$$\begin{aligned} f(x_1 + y_1, x_2 + y_2) &= 2(x_1 + y_1) + 3(x_2 + y_2) = 2x_1 + 2y_1 + 3x_2 + 3y_2 \\ &= [2x_1 + 3x_2] + [2y_1 + 3y_2] = f(x_1, x_2) + f(y_1, y_2). \end{aligned}$$

In addition, for any scalar α, we have

$$f(\alpha x_1, \alpha x_2) = 2\alpha x_1 + 3\alpha x_2 = \alpha[2x_1 + 3x_2] = \alpha f(x_1, x_2).$$

Therefore, this function is indeed linear. This function can also be written as a vector form

$$f(\boldsymbol{x}) = \begin{pmatrix} 2 & 3 \end{pmatrix} \begin{pmatrix} x_1 \\ x_2 \end{pmatrix} = \boldsymbol{a} \cdot \boldsymbol{x} = \boldsymbol{a}^T \boldsymbol{x},$$

where $\boldsymbol{a} \cdot \boldsymbol{x} = \boldsymbol{a}^T \boldsymbol{x}$ is the inner product of $\boldsymbol{a} = (2 \;\; 3)^T$ and $\boldsymbol{x} = (x_1 \;\; x_2)^T$.

In general, functions can be a multiple-component vector, which can be written as $\boldsymbol{F}$ [22]. A function $\boldsymbol{F}$ is called affine if there exists a linear function $\mathcal{L}$ and a vector constant $\boldsymbol{b}$ such that $\boldsymbol{F} = \mathcal{L}(\boldsymbol{x}) + \boldsymbol{b}$. In general, an affine function is a linear function

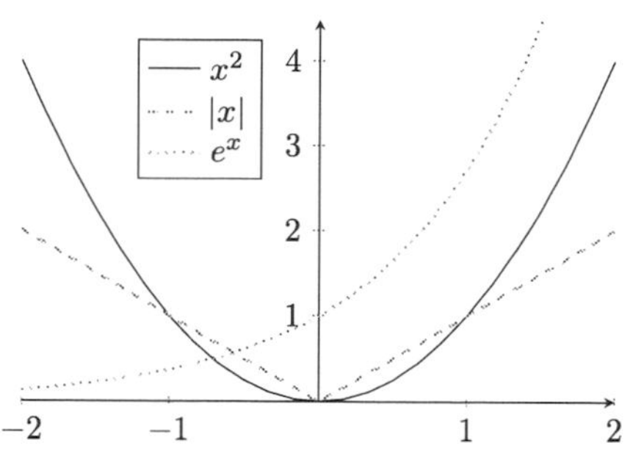

Figure 2.1 Convex functions.

with translation, which can be written in a matrix form $\boldsymbol{F} = \boldsymbol{A}\boldsymbol{x} + \boldsymbol{b}$, where $\boldsymbol{A}$ is an $m \times n$ matrix, and $\boldsymbol{b}$ is a column vector in $\mathbb{R}^n$.

Knowing the properties of a function can be useful for finding the maximum or minimum of the function. In fact, in mathematical optimization, nonlinear problems are often classified according to the convexity of the defining function(s). Geometrically speaking, an object is convex if for any two points within the object, every point on the straight line segment joining them is also within the object. Examples are a solid ball, a cube, and a pyramid. Obviously, a hollow object is not convex.

Mathematically speaking, a set $S \in \mathbb{R}^n$ in a real vector space is called a convex set if

$$\theta \boldsymbol{x} + (1-\theta)\boldsymbol{y} \in S, \qquad \forall(\boldsymbol{x}, \boldsymbol{y}) \in S, \ \theta \in [0, 1]. \tag{2.3}$$

Thus, an affine set is always convex, but a convex set is not necessarily affine.

2.1.2 Convex functions

A function $f(x)$ defined on a convex set Ω is called convex if

$$f(\alpha x + \beta y) \leq \alpha f(x) + \beta f(y), \qquad \forall x, y \in \Omega, \tag{2.4}$$

where

$$\alpha \geq 0, \ \beta \geq 0, \ \alpha + \beta = 1. \tag{2.5}$$

Some examples of convex functions are shown in Fig. 2.1.

Example 8

For example, the convexity of $f(x) = x^2 - 1$ requires

$$(\alpha x + \beta y)^2 - 1 \leq \alpha(x^2 - 1) + \beta(y^2 - 1), \ \forall x, y \in \Re,$$

where $\alpha, \beta \geq 0$ and $\alpha + \beta = 1$. This is equivalent to

$$\alpha x^2 + \beta y^2 - (\alpha x + \beta y)^2 \geq 0,$$

where we have used $\alpha + \beta = 1$. We now have

$$\alpha x^2 + \beta y^2 - \alpha^2 x^2 - 2\alpha\beta xy - \beta^2 y^2$$
$$= \alpha(1-\alpha)(x-y)^2 = \alpha\beta(x-y)^2 \geq 0,$$

which is always true because $\alpha, \beta \geq 0$ and $(x-y)^2 \geq 0$. Therefore, $f(x) = x^2 - 1$ is convex for all $x \in \mathbb{R}$.

A function $f(x)$ on Ω is concave if and only if $g(x) = -f(x)$ is convex. An interesting property of a convex function f is that the vanishing of the gradient $df/dx|_{x_*} = 0$ guarantees that the point x_* is the global minimum of f. Similarly, for a concave function, any local maximum is also the global maximum. If a function is not convex or concave, then it is much more difficult to find its global minimum or maximum.

2.1.3 Mathematical operations on convex functions

There are some important mathematical operations that still preserve the convexity: nonnegative weighted sum, composition using affine functions, and maximization or minimization. For example, if f is convex, then βf is also convex for $\beta \geq 0$. The nonnegative sum $\alpha f_1 + \beta f_2$ is convex if f_1, f_2 are convex and $\alpha, \beta \geq 0$.

The composition using an affine function also holds. For example, $f(\boldsymbol{A}\boldsymbol{x} + \boldsymbol{b})$ is convex if f is convex. In addition, if $f_1, f_2, \ldots, f_n$ are convex, then the maximum of all these functions, $\max\{f_1, f_2, \ldots, f_n\}$, is also convex. Similarly, the piecewise-linear function $\max_{i=1}^{n}(\boldsymbol{A}_i \boldsymbol{x} + \boldsymbol{b}_i)$ is also convex.

If both f and g are convex, then $\psi(\boldsymbol{x}) = f(g(\boldsymbol{x}))$ can also be convex under certain nondecreasing conditions. For example, $\exp[f(\boldsymbol{x})]$ is convex if $f(\boldsymbol{x})$ is convex. This can be extended to the vector composition, and most interestingly, the log-sum-exp function

$$f(\boldsymbol{x}) = \log \sum_{k=1}^{n} e^{x_k}, \tag{2.6}$$

is convex. For a more comprehensive introduction of convex functions, we refer the readers to more advanced literature such as the book by Boyd and Vandenberghe [22].

2.2 Computational complexity

In the description of algorithmic complexity, we often have to use the order notations, often in terms of big O and small o. Loosely speaking, for two functions $f(x)$ and $g(x)$, if

$$\lim_{x \to x_0} \frac{f(x)}{g(x)} \to K, \tag{2.7}$$

where K is a finite, nonzero limit, we write

$$f = O(g). \tag{2.8}$$

The big O notation means that f is asymptotically equivalent to the order of $g(x)$. If the limit is unity or $K = 1$, then we that say $f(x)$ is asymptotically equivalent to $g(x)$. In this particular case, we write

$$f \sim g, \tag{2.9}$$

which is equivalent to $f/g \to 1$ and $g/f \to 1$ as $x \to x_0$. Obviously, x_0 can be any value, including 0 and ∞. The notation $\sim$ does not necessarily mean $\approx$ in general, though it may give the same results, especially in the case where $x \to 0$. For example, $\sin x \sim x$ and $\sin x \approx x$ as $x \to 0$.

When we say f is order of 100 (or $f \sim 100$), this does not mean $f \approx 100$ but rather that f can be between about 50 and 150. The small o notation is often used if the limit tends to 0, that is,

$$\lim_{x \to x_0} \frac{f}{g} \to 0, \tag{2.10}$$

or

$$f = o(g). \tag{2.11}$$

If $g > 0$, then $f = o(g)$ is equivalent to $f \ll g$ (that is, f is much less than g).

Example 9

For example, for all $x \in \mathbb{R}$, we have

$$e^x = 1 + x + \frac{x^2}{2!} + \frac{x^3}{3!} + \cdots + \frac{x^n}{n!} + \cdots, \tag{2.12}$$

which can be written as

$$e^x \approx 1 + x + O(x^2) \approx 1 + x + \frac{x^2}{2} + o(x), \tag{2.13}$$

depending on the accuracy of the approximation of interest.

It is worth pointing out that the expressions in computational complexity are most concerned with functions such as $f(n)$ of an input of problem size n, where $n \in \mathbb{N}$ is an integer in the set of natural numbers $\mathbb{N} = \{1, 2, 3, \ldots\}$.

For example, for the functions $f(n) = 10n^2 + 20n + 100$ and $g(n) = 5n^2$, we have

$$f(n) = O\Big(g(n)\Big) \tag{2.14}$$

for every sufficiently large n. As n is sufficiently large, n^2 is much larger than n (i.e., $n^2 \gg n$), then n^2 terms dominate two expressions. To emphasize the input n, we can often write

$$f(n) = O\Big(g(n)\Big) = O(n^2). \tag{2.15}$$

In addition, $f(n)$ is in general a polynomial of n, which not only includes terms such as n^3 and n^2, but it also may include $n^{2.5}$ or $\log(n)$. Therefore, $f(n) = 100n^3 + 20n^{2.5} + 25n\log(n) + 123n$ is a valid polynomial in the context of computational complexity. In this case, we have

$$f(n) = 100n^3 + 20n^{2.5} + 25n\log(n) + 123n = O(n^3). \tag{2.16}$$

Here, we always implicitly assume that n is sufficiently large and the base of the logarithm is 2.

To measure how easily or hardly a problem can be solved, we need to estimate its computational complexity. We cannot simply ask how long it takes to solve a particular problem instance because the actual computational time depends on both hardware and software used to solve it. Thus, time does not make much sense in this context. A useful measure of complexity should be independent of the hardware and software used. However, such complexity is closely linked to the algorithms used.

2.2.1 Time and space complexity

To find the maximum (or minimum) among n different numbers, we only need to go through each number once by simply comparing the current number with the highest (or lowest) number once and update the new highest (or lowest) when necessary. Thus, the number of mathematical operations is simply $O(n)$, which is the time complexity of this problem.

In practice, comparing two big numbers may take slightly longer, and different representations of numbers can also affect the speed of this comparison. In addition, multiplication and division usually take more time than simple addition and substraction. However, in computational complexity, we usually ignore such minor differences and simply treat all operations as equal. In this sense, the complexity is about the number or order of mathematical operations, not the actual order of computational time.

On the other hand, space computational complexity estimates the size of computer memory needed to solve the problem. In the previous simple problem of finding the maximum or minimum among n different numbers, the memory needed is $O(n)$ because it needs to store n different numbers at n different entries in the computer memory. Though we need one more entry to store the largest or smallest number, this minor change does not affect the order of complexity because we implicitly assume that n is sufficiently large [6,58].

In most literature, if there is no time or space explicitly used when talking about computational complexity, it usually means time complexity. In discussing computational complexity, we often use the word "problem" to mean a class of problems of the same type and an "instance" to mean a specific example of a problem class. Thus, $\boldsymbol{Ax} = \boldsymbol{b}$ is a problem (class) for linear algebra, whereas

$$\begin{pmatrix} 2 & 3 \\ 1 & 1 \end{pmatrix} \begin{pmatrix} x \\ y \end{pmatrix} = \begin{pmatrix} 8 \\ 3 \end{pmatrix} \tag{2.17}$$

is an instance. In addition, a decision problem is a yes–no problem where an output is binary (0 or 1), even though the inputs can be any values.

The computational complexity is closely linked to the type of problems. For the same type of problems, different algorithms can be used, and the number of basic mathematical operations may be different. In this case, we are concerned with the complexity of an algorithm in terms of arithmetic complexity.

2.2.2 Complexity of algorithms

The computational complexity discussed up to now has focused on the problems, and the algorithms are mainly described simply in terms of polynomial or exponential time. From the perspective of algorithm development and analysis, different algorithms will have different complexity even for the same type of problems. In this case, we have to estimate the arithmetic complexity of an algorithm or simply algorithmic complexity.

For example, to solve a sorting problem with n different numbers so as to sort them from the smallest to the largest, we can use different algorithms. For example, the selection sort uses two loops for sorting n, which has an algorithmic complexity of $O(n^2)$, whereas the quicksort (or partition and exchange sort) has a complexity of $O(n \log n)$. There are many different sorting algorithms with different complexities.

It is worth pointing out that the algorithmic complexity here is mainly about time complexity because the space (memory) complexity is less important. In this case, the space algorithmic complexity is $O(n)$.

Example 10

The multiplication of two $n \times n$ matrices $\boldsymbol{A}$ and $\boldsymbol{B}$ using simple matrix multiplication rules has a complexity of $O(n^3)$. There are n rows and n columns for each matrix, and their product $\boldsymbol{C}$ has $n \times n$ entries. To get each entry, we need to carry out the multiplication of a row of $\boldsymbol{A}$ by a corresponding column of $\boldsymbol{B}$ and calculate their sum, and thus the complexity is $O(n)$. As there are $n \times n = n^2$ entries, the overall complexity is $O(n^2) \times O(n) = O(n^3)$.

In the rest of this book, we analyze different algorithms; the complexity to be given is usually the arithmetic complexity of an algorithm under discussion.

2.3 Norms and regularization

2.3.1 Norms

In general, a vector in an n-dimensional space ($n \geq 1$) can be written as a column vector

$$\boldsymbol{x} = \begin{pmatrix} x_1 \\ x_2 \\ \vdots \\ x_n \end{pmatrix} = (x_1, x_2, \ldots, x_n)^T \tag{2.18}$$

or a row vector

$$\boldsymbol{x} = \begin{pmatrix} x_1 & x_2 & \ldots & x_n \end{pmatrix}. \tag{2.19}$$

A simple transpose (T) can convert a column vector into its corresponding row vector. The length of $\boldsymbol{x}$ can be written as

$$||\boldsymbol{x}|| = \sqrt{x_1^2 + x_2^2 + \cdots + x_n^2}, \tag{2.20}$$

which is the Euclidean norm.

The addition or substraction of two vectors $\boldsymbol{u}$ and $\boldsymbol{v}$ are the addition or substraction of their corresponding components, that is,

$$\boldsymbol{u} \pm \boldsymbol{v} = \begin{pmatrix} u_1 \\ u_2 \\ \vdots \\ u_n \end{pmatrix} \pm \begin{pmatrix} v_1 \\ v_2 \\ \vdots \\ v_n \end{pmatrix} = \begin{pmatrix} u_1 \pm v_1 \\ u_2 \pm v_2 \\ \vdots \\ u_n \pm v_n \end{pmatrix}. \tag{2.21}$$

The dot product, also called the inner product, of two vectors $\boldsymbol{u}$ and $\boldsymbol{v}$ is defined as

$$\boldsymbol{u}^T \boldsymbol{v} \equiv \boldsymbol{u} \cdot \boldsymbol{v} = \sum_{i=1}^{n} u_i v_i = u_1 v_1 + u_2 v_2 + \cdots + u_n v_n. \tag{2.22}$$

For an n-dimensional vector $\boldsymbol{x}$, we can define the p-norm or L_p-norm (also L^p-norm) as

$$||\boldsymbol{x}||_p \equiv \Big(|x_1|^p + |x_2|^p + \cdots + |x_n|^p\Big)^{1/p} = \Big(\sum_{i=1}^{n} |x_i|^p\Big)^{1/p}, \quad p > 0. \tag{2.23}$$

Obviously, the Cartesian norm or length is the L_2-norm

$$||\boldsymbol{x}||_2 = \sqrt{|x_1|^2 + |x_2|^2 + \cdots + |x_n|^2} = \sqrt{x_1^2 + x_2^2 + \cdots + x_n^2}. \tag{2.24}$$

Three most widely used norms are $p = 1, 2$, and ∞ [160]. When $p = 2$, it becomes the Cartesian L_2-norm as discussed before. When $p = 1$, the L_1-norm is given by

$$||\boldsymbol{x}||_1 = |x_1| + |x_2| + \cdots + |x_n|. \tag{2.25}$$

For $p = \infty$, it becomes

$$||\boldsymbol{x}||_\infty = \max\{|x_1|, |x_2|, \ldots, |x_n|\} = x_{\max}, \tag{2.26}$$

which is the largest absolute component of $\boldsymbol{x}$. This is because

$$\begin{aligned} ||\boldsymbol{x}||_\infty &= \lim_{p\to\infty} \Big(\sum_{i=1}^{p} |x_i|^p\Big)^{1/p} = \lim_{p\to\infty} \Big(|x_{\max}|^p \sum_{i=1}^{n} \Big|\frac{x_i}{x_{\max}}\Big|^p\Big)^{1/p} \\ &= x_{\max} \lim_{p\to\infty} \Big(\sum_{i=1}^{n} \Big|\frac{x_i}{x_{\max}}\Big|\Big)^{1/p} = x_{\max}, \end{aligned} \tag{2.27}$$

where we have used the fact that $|x_i/x_{\max}| < 1$ (except for one component, say, $|x_k| = x_{\max}$). Thus, $\lim_{p\to\infty} |x_i/x_{\max}|^p \to 0$ for all $i \neq k$. Thus, the sum of all ratio terms is 1, that is,

$$\Big(\lim_{p\to\infty} \Big|\frac{x_i}{x_{\max}}\Big|^p\Big)^{1/p} = 1. \tag{2.28}$$

In general, for any two vectors $\boldsymbol{u}$ and $\boldsymbol{v}$ in the same space, we have the inequality

$$||\boldsymbol{u}||_p + ||\boldsymbol{v}||_p \geq ||\boldsymbol{u} + \boldsymbol{v}||_p, \quad p \geq 0. \tag{2.29}$$

Example 11

For two vectors $\boldsymbol{u} = [1 \;\; 2 \;\; 3]^T$ and $\boldsymbol{v} = [1 \;\; -2 \;\; -1]^T$, we have

$$\boldsymbol{u}^T\boldsymbol{v} = 1 \times 1 + 2 \times (-2) + 3 \times (-1) = -6,$$

$$||\boldsymbol{u}||_1 = |1| + |2| + |3| = 6, \quad ||\boldsymbol{v}||_1 = |1| + |-2| + |-1| = 4,$$

$$||\boldsymbol{u}||_2 = \sqrt{1^2 + 2^2 + 3^2} = \sqrt{14}, \quad ||\boldsymbol{v}||_2 = \sqrt{1^2 + (-2)^2 + (-1)^2} = \sqrt{6},$$

$$||\boldsymbol{u}||_\infty = \max\{|1|, |2|, |3|\} = 3, \quad ||\boldsymbol{v}||_\infty = \max\{|1|, |-2|, |-1|\} = 2,$$

and

$$\boldsymbol{w} = \boldsymbol{u} + \boldsymbol{v} = \begin{pmatrix} 1+1 & 2+(-2) & 3+(-1) \end{pmatrix}^T = \begin{pmatrix} 2 & 0 & 2 \end{pmatrix}^T$$

with norms

$$||\boldsymbol{w}||_1 = |2| + |0| + |2| = 2, \quad ||\boldsymbol{w}||_\infty = \max\{|2|, |0|, |2|\} = 2,$$

$$||\boldsymbol{w}||_2 = \sqrt{2^2 + 0^2 + 2^2} = \sqrt{8}.$$

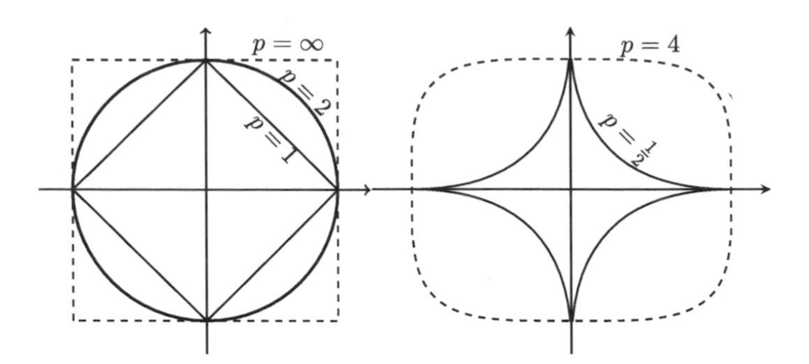

Figure 2.2 Different p-norms for $p = 1, 2,$ and ∞ (left) as well as $p = 1/2$ and $p = 4$ (right).

Using these values, it is straightforward to verify that

$$||\boldsymbol{u}||_p + ||\boldsymbol{v}||_p \geq ||\boldsymbol{u} + \boldsymbol{v}||_p \quad (p = 1, 2, \infty).$$

In the particular case of two-dimensional (2D) vectors, different norms $L_p = (|x|^p + |y|^p)^{1/p}$ with different values of p are shown in Fig. 2.2.

2.3.2 Regularization

In many applications such as curve-fitting and machine learning, overfitting can be a serious issue, and one way to avoid overfitting is using regularization. Loosely speaking, regularization is using some penalty term added to the objective or loss function so as to constrain certain model parameters. For example, in the method of least-squares and many learning algorithms, the objective is to minimize the loss function $L(x)$, which represents the errors between data labels y_i and the predictions $f_i = f(x_i)$ for m data points (x_i, y_i), $i = 1, 2, \ldots, m$, that is,

$$L(x) = \sum_{i=1}^{m} \big[y_i - f(x_i)\big]^2, \tag{2.30}$$

which is the L_2-norm of the errors $E_i = y_i - f_i$. The model prediction $f(x, \boldsymbol{w})$ usually have many model parameters such as $\boldsymbol{w} = (w_1, w_2, ..., w_K)$ for simple polynomial curve-fitting. In general, a prediction model can have K different model parameters, overfitting can occur if the model becomes too complex with too many parameters, and the oscillations become significant. Thus, a penalty term in terms of some norm of the model parameters is usually added to the loss function. For example, the well-known Tikhonov regularization uses the L_2-norm, and we have

$$\text{minimize} \sum_{i=1}^{m} \Big[y_i - f(x_i, \boldsymbol{w})\Big]^2 + \lambda||\boldsymbol{w}||_2, \tag{2.31}$$

where $\lambda > 0$ is the penalty parameter. Obviously, other norms can be used. For example, in the Lasso method, the regularization uses 1-norm, which gives

$$\text{minimize } \frac{1}{m}\sum_{i=1}^{m}\Big[y_i - f(x_i, \boldsymbol{w})\Big]^2 + \lambda||\boldsymbol{w}||_1. \tag{2.32}$$

We will introduce both the method of least-squares and Lasso method in late chapters.

2.4 Probability distributions

2.4.1 Random variables

For a discrete random variable X with distinct values such as the number of cars passing through a junction, each value x_i may occur with certain probability $p(x_i)$. In other words, the probability varies and is associated with the corresponding random variable. Traditionally, an uppercase letter such as X is used to denote a random variable, whereas a lowercase letter such as x_i represents its values. For example, if X means a coin-flipping event, then $x_i = 0$ (tail) or 1 (head). A probability function $p(x_i)$ is a function that assigns probabilities to all the discrete values x_i of the random variable X.

As an event must occur inside a sample space, the requirement that all the probabilities must be summed to one, which leads to

$$\sum_{i=1}^{n} p(x_i) = 1. \tag{2.33}$$

For example, the outcomes of tossing a fair coin form a sample space. The outcome of a head (H) is an event with probability $P(H) = 1/2$, and the outcome of a tail (T) is also an event with probability $P(T) = 1/2$. The sum of both probabilities should be one, that is,

$$P(H) + P(T) = \frac{1}{2} + \frac{1}{2} = 1. \tag{2.34}$$

The cumulative probability function of X is defined by

$$P(X \le x) = \sum_{x_i < x} p(x_i). \tag{2.35}$$

Two main measures for a random variable X with given probability distribution $p(x)$ are its mean and variance. The mean μ or expectation of $E[X]$ is defined by

$$\mu \equiv E[X] \equiv <X> = \int x p(x) dx \tag{2.36}$$

for a continuous distribution and the integration is within the integration limits. If the random variable is discrete, then the integration becomes the weighted sum

$$E[X] = \sum_i x_i p(x_i). \tag{2.37}$$

The variance $\text{var}[X] = \sigma^2$ is the expectation value of the deviation squared, that is, $E[(X-\mu)^2]$. We have

$$\sigma^2 \equiv \text{var}[X] = E[(X-\mu)^2] = \int (x-\mu)^2 p(x)dx. \tag{2.38}$$

The square root of the variance $\sigma = \sqrt{\text{var}[X]}$ is called the standard deviation, which is simply σ.

The above definition of mean $\mu = E[X]$ is essentially the first moment if we define the kth moment of a random variable X (with a probability density distribution $p(x)$) by

$$\mu_k \equiv E[X^k] = \int x^k p(x)\text{d}x \quad (k = 1, 2, 3, \ldots). \tag{2.39}$$

Similarly, we can define the kth central moment by

$$\begin{aligned}\nu_k &\equiv E[(X - E[X])^k] \equiv E[(X-\mu)^k] \\ &= \int (x-\mu)^k p(x)\text{d}x \quad (k = 0, 1, 2, 3, \ldots),\end{aligned} \tag{2.40}$$

where μ is the mean (the first moment). Thus, the zeroth central moment is the sum of all probabilities when $k = 0$, which gives $\nu_0 = 1$. The first central moment is $\nu_1 = 0$. The second central moment ν_2 is the variance σ^2, that is, $\nu_2 = \sigma^2$.

2.4.2 Probability distributions

There are a number of other important distributions such as the normal distribution, Poisson distribution, exponential distribution, binomial distribution, Cauchy distribution, Lévy distribution, and Student t-distribution.

A Bernoulli distribution is a distribution of outcomes of a binary random variable X where the random variable can only take two values, either 1 (success or yes) or 0 (failure or no). The probability of taking 1 is $0 \le p \le 1$, whereas the probability of taking 0 is $q = 1 - p$. Then, the probability mass function can be written as

$$B(m, p) = \begin{cases} p & \text{if } m = 1, \\ 1 - p, & \text{if } m = 0, \end{cases} \tag{2.41}$$

which can be written more compactly as

$$B(m, p) = p^m (1-p)^{1-m}, \quad m \in \{0, 1\}. \tag{2.42}$$

It is straightforward to show that its mean and variance are

$$E[X] = p, \quad \text{var}[X] = pq = p(1-p). \tag{2.43}$$

This is the probability of a single experiment with two distinct outcomes. In the case of multiple experiments or trials (n), the probability distribution of exactly m successes becomes the binomial distribution

$$B_n(m, n, p) = \binom{n}{m} p^m (1-p)^{n-m}, \tag{2.44}$$

where

$$\binom{n}{m} = \frac{n!}{m!(n-m)!} \tag{2.45}$$

is the binomial coefficient. Here, $n!$ is the factorial, $n! = n(n-1)(n-2)\ldots 1$. For example, $5! = 5 \times 4 \times 3 \times 2 \times 1 = 120$. Conventionally, we set $0! = 1$.

It is also straightforward to verify that

$$E[X] = np, \quad \text{var}[X] = np(1-p) \tag{2.46}$$

for n trials.

The exponential distribution has the probability density function

$$f(x) = \lambda e^{-\lambda x}, \quad \lambda > 0 \quad (x > 0), \tag{2.47}$$

and $f(x) = 0$ for $x \leq 0$. Its mean and variance are

$$\mu = 1/\lambda, \qquad \sigma^2 = 1/\lambda^2. \tag{2.48}$$

The Poisson distribution is the distribution for small-probability discrete events. Typically, it is concerned with the number of events that occur in a certain time interval (e.g., the number of telephone calls in an hour) or spatial area.

The probability density function of the Poisson distribution is

$$P(X = x) = \frac{\lambda^x e^{-\lambda}}{x!}, \qquad \lambda > 0, \tag{2.49}$$

where $x = 0, 1, 2, \ldots, n$, and λ is the mean of the distribution.

The Gaussian distribution or normal distribution is the most important continuous distribution in probability, and it has a wide range of applications. For a continuous random variable X, the probability density function (PDF) of the Gaussian distribution is given by

$$p(x) = \frac{1}{\sigma\sqrt{2\pi}} e^{-\frac{(x-\mu)^2}{2\sigma^2}}, \tag{2.50}$$

where $\sigma^2 = \text{var}[X]$ is the variance, and $\mu = E[X]$ is the mean of the Gaussian distribution. The total probability of unity requires that

$$\int_{-\infty}^{\infty} p(x)dx = 1, \tag{2.51}$$

and this is exactly the reason why the factor $1/\sqrt{2\pi}$ is required in the normalization of all the probabilities.

Cauchy probability distribution can be written as

$$p(x, \mu, \gamma) = \frac{1}{\pi\gamma}\Big[\frac{\gamma^2}{(x-\mu)^2 + \gamma^2}\Big], \quad -\infty < x < \infty; \tag{2.52}$$

its mean and variance are undefined or infinite, which is a true indication that this distribution is heavy-tailed. The cumulative distribution function of the Cauchy distribution is

$$F(x) = \frac{1}{\pi}\tan^{-1}\Big(\frac{x-\mu}{\gamma}\Big) + \frac{1}{2}. \tag{2.53}$$

It is worth pointing out that this distribution can have a heavy tail or a fat tail where probability can be still significantly nonzero at the tails as $x \to \infty$. Thus, such a distribution belongs to the heavy-tailed or fat-tailed distributions.

Other heavy-tailed distributions include Pareto distribution, power-law distributions, and Lévy distribution.

2.4.3 Conditional probability and Bayesian rule

In the above calculations of probabilities, we have assumed that all possible outcomes of an experiment such as tossing a coin are independent of each other, and events are independent of each other. In general, two events A and B are independent if the events have no influence on each other. That is to say, the occurrence of event A does not affect or provide any information about whether event B will occur. In this case, the probability of both events occurring (their joint probability) is denoted as $P(A \cap B)$ and is simply the product of the probabilities of each individual event, $P(A)$ and $P(B)$, respectively. We have

$$P(A \cap B) = P(A)P(B). \tag{2.54}$$

Probabilities may change if additional information is known or some other event has already occurred. In this case, we are dealing with conditional probabilities. Let $P(B|A)$ denote the probability that the event B will occur given that event A has already occurred. Here, "$B|A$" means that the event B occurs give that event A has occurred, and the outcome of event A can be considered as data or evidence.

As the events A and B are no long independent, their joint probability is becomes

$$P(A \cap B) = P(A)P(B|A) = P(B|A)P(A), \tag{2.55}$$

which is often called the multiplication rule in probability. Similarity, we can also have

$$P(A \cap B) = P(B \cap A) = P(B)P(A|B) = P(A|B)P(B). \tag{2.56}$$

Thus, the conditional probability $P(B|A)$ can be calculated by

$$P(B|A) = \frac{P(B \cap A)}{P(A)} = \frac{P(B)P(A|B)}{P(A)}, \tag{2.57}$$

which is the well-known Bayes' theorem or Bayesian rule. As the whole event space Ω can be decomposed as $\Omega = B \cup \bar{B}$ (the union of event B and event $\bar{B} =$ not B), the total probability should be one, that is, $P(\bar{B}) = 1 - P(B)$, so we can calculate $P(A)$ by

$$P(A) = P(B)P(A|B) + P(\bar{B})P(A|\bar{B}), \tag{2.58}$$

which can be generalized to the sum of all possible events. Thus the Bayes theorem becomes

$$P(B|A) = \frac{P(B)P(A|B)}{P(A)} = \frac{P(B)P(A|B)}{P(B)P(A|B) + P(\bar{B})P(A|\bar{B})}. \tag{2.59}$$

The Bayes theorem can be a useful tool for many applications.

Example 12

Consider a hypothetical example for drug tests in sports. It is believed that a particular method of drug testing can have an accuracy of 99% if athletes are taking drugs. For athletes not taking drugs, the positive test is only 0.5%. In a particular sport event, it is assumed that only 1 of 1000 athletes takes this kind of drug. Now suppose an athlete is selected at random and the test shows positive for the drug. What is the probability that the athlete is taking the drug?

Let B denote the event that an athlete is taking the drug, and let A denote the event that the individual test is positive. We have $P(B) = 1/1000 = 0.001$, $P(A|B) = 0.99$, and $P(A|\bar{B}) = 0.005$. Thus, the probability that the athlete is actually taking the drug is

$$\begin{aligned} P(B|A) &= \frac{P(B)P(A|B)}{P(B)P(A|B) + P(\bar{B})P(A|\bar{B})} \\ &= \frac{0.001 \times 0.99}{0.001 \times 0.99 + 0.999 \times 0.005} \approx 0.165, \end{aligned} \tag{2.60}$$

which is surprisingly a low probability.

In a more general sense, we have a set of observed data $y_1, y_2, \ldots, y_m$ for a random variable y where y_i can be an n-dimensional vector itself. Though we may not know exactly the probability density function $p(y)$ to generate such a data set, however, we may wish to provide some insight into such distributions.

Suppose we have a family of conditional probability function for y described by a parameter β, which itself is a random variable, and in general β can be a vector of many different parameters. Thus, we use $p(y|\beta)$ to denote this family of conditional probability distributions, which is often referred to as the likelihood function. If we have some knowledge about β and its corresponding model for generating the given data, then we can use $p(\beta)$ to denote the prior distribution over β [51].

Now we can use the Bayesian rule to estimate the conditional probability function $p(\beta|y)$ for β given the data; we have

$$p(\beta|y) = \frac{p(y|\beta)p(\beta)}{p(y)}, \tag{2.61}$$

which is called the posterior distribution. In essence, this provides a posterior estimate for a distribution, based on the observed data and the prior distribution.

It is worth pointing out that the data model or distribution $p(y)$ does not contain the model parameter β, which is in fact the sum or integration over all parameter values, that is,

$$p(y) = \int p(y|\tilde{\beta})p(\tilde{\beta})d\tilde{\beta}, \tag{2.62}$$

but this integral is usually very difficult to calculate, if not impossible. In most cases, Markov chain Monte Carlo numerical simulations can be used to get a good estimate. We will briefly introduce some of these numerical sampling methods later.

2.4.4 Gaussian process

Loosely speaking, a Gaussian process is a continuous stochastic process for a set of random variables $\boldsymbol{X} = [X_1, X_2, \ldots, X_m]^T$, and any such finite set obeys a multivariate normal distribution with probability density function

$$p(x_1, x_2, \ldots, x_m) = \frac{1}{\sqrt{(2\pi)^m |\boldsymbol{\Sigma}|}} \exp\Big[-\frac{1}{2}(\boldsymbol{x}-\boldsymbol{\mu})\boldsymbol{\Sigma}^{-1}(\boldsymbol{x}-\boldsymbol{\mu})\Big], \tag{2.63}$$

where $\boldsymbol{\mu}$ is the mean vector of m individual variables

$$\begin{aligned}\boldsymbol{\mu} &\equiv E[\boldsymbol{X}] = [E[X_1], \ldots, E[X_m]]^T \\ &= [\mu_1, \ldots, \mu_m]^T,\end{aligned} \tag{2.64}$$

and $\boldsymbol{\Sigma}$ is the covariance matrix of size $m \times m$:

$$\boldsymbol{\Sigma} = E\big[(\boldsymbol{X}-\boldsymbol{\mu})(\boldsymbol{X}-\boldsymbol{\mu})^T\big] = \begin{pmatrix} \text{cov}(X_1, X_1) & \ldots & \text{cov}(X_1, X_m) \\ \text{cov}(X_2, X_1) & \ldots & \text{cov}(X_2, X_m) \\ \vdots & \ddots & \vdots \\ \text{cov}(X_m, X_1) & \ldots & \text{cov}(X_m, X_m) \end{pmatrix}$$

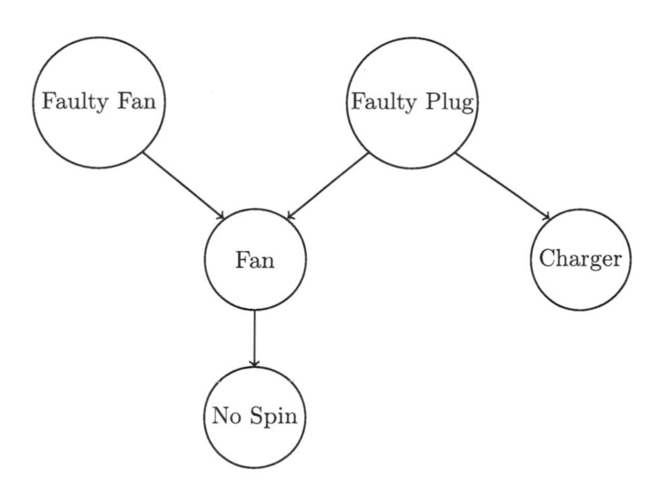

Figure 2.3 A simple Bayesian network example.

$$-\begin{pmatrix} E[(X_1-\mu_1)(X_1-\mu_1)] & \dots & E[(X_1-\mu_1)(X_m-\mu_m)] \\ E[(X_2-\mu_2)(X_1-\mu_1)] & \dots & E[(X_2-\mu_2)(X_m-\mu_m)] \\ \vdots & \ddots & \vdots \\ E[(X_m-\mu_m)(X_1-\mu_1)] & \dots & E[(X_m-\mu_m)(X_m-\mu_m)] \end{pmatrix}. \tag{2.65}$$

In addition, $\mathbf{\Sigma}^{-1}$ and $|\mathbf{\Sigma}| = \det(\Sigma)$ are the inverse and the determinant of the matrix Σ.

Furthermore, any linear combination of these variables obeys a normal distribution. Gaussian processes are important to many applications such as Bayesian inference and machine learning [121]. In fact, Gaussian processes are the foundation for kernel-based methods or kernel machines such as support vector machines and least-squares classifiers. Their main differences are the probabilistic point of view and perspectives, in addition to some key technical details. Many Bayesian approaches use Gaussian distributions as their prior distributions; they form a class of methods such as Kriging, which is a Gaussian process regression technique. The interested readers can refer to more specialized literature [121].

2.5 Bayesian network and Markov models

A Bayesian network (BN) is a probabilistic graphical model for representing knowledge about an uncertain domain where each node corresponds to a random variable and each edge represents the conditional probability for the corresponding random variables [9]. BNs are also called belief networks or Bayes nets. Due to dependencies and conditional probabilities, a BN corresponds to a directed acyclic graph (DAG) where no loop or self connection is allowed. For example, Fig. 2.3 is a BN.

Let us use an example to show how it works. This example is very similar to the "earthquake" example by Pearl [112] and the "chair" example by Ben-Gal [9]. In my office, there is an electric fan that I use often in summer and not in other seasons. Imagine a scenario that I try to switch on the fan, but it does not spin. The fan is plugged into an extension socket or plug, and there is a possibility of a plug failure. How do we figure out what the possible causes are?

The fan has a probability of 0.02 for failure, whereas the plug is very old and has the failure probability 0.2. I also have a mobile phone charger connected to the same plug. I found that the charger works well. What is the probability of the problem caused by a faulty fan?

We can represent this scenario as a simple Bayesian network, shown in Fig. 2.3. In this case, the parents of the random variable Fan are the nodes Faulty Fan and Faulty Plug, whereas the child of Fan is No Spin. The two variables Faulty Fan and Faulty Plug are marginally independent; however, they become conditionally dependent, given Fan.

The number required to completely specify the probability distributions for a network can be huge. For a set of n binary random variables, it requires $2^n - 1$ joint probability distributions [29]. Even for a small $n = 20$, this number becomes $2^{20} - 1 = 1048575$, which is a huge number. Thus, the complete specification and the exact solution, if possible, can be NP-hard. Therefore, approximate solutions are often sought in practice, and Monte Carlo simulations can be very useful in this case.

Though BNs are directed acyclic graphs of graphical models, however, probabilistic graphical models can have undirected edges, which become the Markov networks or Markov random fields. It is worth pointing out that many conventional machine learning techniques, such as artificial neural networks, Kalman filters, and hidden Markov models can all be considered as particular cases of Bayesian networks as pointed out by Gen-Gal et al. [9].

Bayesian networks have a diverse range of applications [9,29,84,106], and Bayesian statistics is relevant to modern techniques in data mining and machine learning [106–108]. The interested readers can refer to more specialized literature on information theory and learning algorithms [98] and Bayesian approach for neural networks [91].

2.6 Monte Carlo sampling

One of the main difficulties in Bayesian statistics is to estimate complex probability distributions, and their integrals for normalization can become intractable. Monte Carlo-based methods can be a powerful technique to draw samples and thus estimate such distributions very accurately [48,50]. There are many different Monte Carlo methods, but we will mainly focus on the Markov chain Monte Carlo (MCMC) methods. More specifically, we will introduce the Gibbs sampler and Metropolis–Hastings algorithms.

2.6.1 Markov chain Monte Carlo

A Markov chain is a sequence of random variables $X_1, X_2, \ldots$ with proper Markov properties, where the next state depends on the current state and the transition probability, that is, the conditional probability

$$\begin{aligned} &P(X_{k+1}=s|X_k=s_k,\ldots,X_2=s_2,X_1=s_1) \\ &\quad = P(X_{k+1}=s|X_k=s_k), \end{aligned} \tag{2.66}$$

does not depend on the states or values before k.

Such chains can have stable probability distributions as $k \to \infty$, which will forget their initial states. Suppose we wish to estimate the states of a random parameter θ; two important things are the current values θ and the transition probability or transition kernel

$$T(\theta,\theta') = p_{\theta\to\theta'}(\theta'|\theta) \tag{2.67}$$

from θ to a new state θ'. Under certain symmetrical balance conditions, there exists a stationary distribution $\pi(\theta)$ for a Markov chain with this transition kernel $T(\theta,\theta')$. It requires that

$$T(\theta,\theta')\pi(\theta) = T(\theta',\theta)\pi(\theta') \tag{2.68}$$

for all θ, θ'.

The key idea of any MCMC approach is constructing a Markov chain with proper Markov properties such that its stationary distribution π is the same as the posterior distribution $p(\theta|y)$ that we wish to estimate. To achieve this, we need a proposal distribution or a jump distribution $q(\theta,\theta')$ so as to propose a new θ' from θ. After the proposed move, a certain criterion is evaluated so as to decide to accept or reject the move from θ to θ'.

There many different algorithms and variants of MCMC methods. We will discuss two of the most widely used: Metropolis–Hastings algorithm [102,103,63] and Gibbs sampler [52].

2.6.2 Metropolis–Hastings algorithm

The main idea of Metropolis–Hastings algorithm can be described in the following steps: candidate proposal, criterion evaluation, and move update.

Suppose we draw a distribution $p(\theta)$, which can be a very complicated probability distribution. We choose a proposal distribution $q(\cdot,\cdot)$ such as Gaussian or uniform distributions. The choice should allow nonzero probability for any possible value of θ. Starting with a starting point $\theta=\theta_0$ at iteration $k=1$, we can draw a candidate θ_* from the proposal distribution $q(\theta_{k-1},\cdot)$. Then we calculate the ratio

$$r = \min\left\{\frac{p(\theta_*)q(\theta_*, \theta_{k-1})}{p(\theta_{k-1})q(\theta_{k-1}, \theta_*)}, 1\right\}. \tag{2.69}$$

The move from θ_{k-1} to θ_* is accepted with probability r; otherwise, the move is discarded [103,63]. If the ratio of the first term on the right is greater than 1, then $r = 1$, which means that it always accepts the move. In general, it accepts with a probability or a fraction of the number of moves in practice.

The key advantage of using the ratio r is that it can deal with complex probability distributions even though we do not know how to calculate the normalization integral constant. Because whatever the integral may be, it is just a constant, and it will disappear from the ratio. Thus, it does not matter anyway.

This iterative process of Metropolis–Hastings (MH) algorithm can be summarized as Algorithm 1.

Algorithm 1 Metropolis–Hastings algorithm.

1: Initial guess θ_0 at $k = 1$
2: **for** (a given number of samples) **do**
3: Propose a candidate θ_* by drawing from a proposal distribution
4: Calculate r using Eq. (2.69)
5: Accept the move $\theta_k \leftarrow \theta_*$ with probability r
6: Otherwise, reject the move and set $\theta_k \leftarrow \theta_{k-1}$ (no change)
7: **end for**

In a very particular case where the proposal distribution is symmetric such that $q(a, b) = q(b, a)$, then Eq. (2.69) becomes

$$r = \min\left\{\frac{p(\theta_*)}{p(\theta_{k-1})}, 1\right\}. \tag{2.70}$$

The Metropolis–Hastings algorithm becomes the classic Metropolis algorithm, where the jump probability density is symmetric. In this case, if the probability $p(\theta_*)$ is higher than the old probability $p(\theta_{k-1})$, and thus their ratio $p(\theta_*)/p(\theta_{k-1})$ is greater than 1, then $r = 1$ from Eq. (2.70), which accepts the moves. This is consistent with the observations that a move to a higher probability region should be always accepted [102,103].

It is worth pointing out that the actual samples drawn from this will usually converge well toward the desired distribution $p(\theta)$. However, the initial part of the sequences may not strictly obey the final stationary distribution, and in practice a burn-in period is used by discarding a fixed number of initial samples in the drawn sequence.

Clearly, another important issue is the choice of the jump distribution. We have to be able to draw samples from it quite quickly. If the jump is too far, then many moves will be rejected. If the jump is too small, the samples may be aggregates at certain part of the distribution, and thus the samples become biased. Some proper balance and tuning may be needed.

2.6.3 Gibbs sampler

The well-known Gibbs sampler, developed by Geman and Geman in 1984 [52], can be considered as a particular case of the MH algorithm by setting $r = 1$. It can be much faster for some special applications.

The main idea is to deal with a multivariate distribution by considering it as a univariate conditional distribution at any time, using the values already drawn for other variables. The sequence is generated by drawing samples from the conditional probability $p(\theta^{(i)}|\ldots)$ for each dimension $i = 1, 2, \ldots, n$.

There are some extensive studies concerning the Gibbs sampler an its variants. There are also very good software packages such as BUGS.

2.7 Entropy, cross entropy, and KL divergence

2.7.1 Entropy and cross entropy

Entropy is an important concept in statistical mechanics, which is also introduced in information theory. The classic entropy S is defined as

$$H = -k_B \sum_i p_i \log p_i, \tag{2.71}$$

where p_i is the probability in state i, and k_B the Boltzmann constant. However, in information theory, the Shannon entropy H is defined by

$$H(p) = -\sum_i p(x_i) \log[p(x_i)], \tag{2.72}$$

where log is usually in the base $b = 2$. This is information averaged over all the possible values x_i of a random variable X. Shannon's information quantity is defined as

$$I(p) = -\log p, \tag{2.73}$$

where the base b of the logarithm can be 2, e, or 10. Different bases only lead to a fixed factor for different units. For $b = 2$, e, and 10, the units are bit, nat, and hartley, respectively. Thus, in practice, we can simply write "log" without worrying the actual base, and we can use any base of convenience. So, the Shannon entropy $H(p) \geq 0$ can be considered as the average amount of information.

In the case of continuous variable, the summation becomes the integral

$$H(p) = \int p(x) \log[1/p(x)] \mathrm{d}x = -\int p(x) \log[p(x)] \mathrm{d}x, \tag{2.74}$$

which integrates over the whole domain of the probability function $p(x)$.

Example 13

For events obeying an exponential distribution

$$p(x) = \lambda e^{-\lambda x}, \quad x \in [0, \infty), \quad \lambda > 0, \tag{2.75}$$

the corresponding entropy can be calculated by

$$\begin{aligned} H(p) &= -\int p(x) \log p(x) dx = -\int_0^{\infty} (\lambda e^{-\lambda x})[-\lambda x + \ln(\lambda)] dx \\ &= \lambda^2 \int_0^{\infty} x e^{-\lambda x} dx - \lambda \ln(\lambda) \int_0^{-\infty} dx = 1 - \ln(\lambda), \end{aligned} \tag{2.76}$$

where we have used

$$\int_0^{\infty} x e^{-\lambda x} dx = \frac{1}{\lambda^2}, \quad \int_0^{\infty} e^{-\lambda x} dx = \frac{1}{\lambda}. \tag{2.77}$$

A very important related concept is the cross entropy $H(p, q)$ of $p(x)$ and $q(x)$. We have

$$H(p, q) = -\int p(x) \log[q(x)] dx, \tag{2.78}$$

which is in essence the average or expectation of $q(x)$ over the probability density function $p(x)$. Alternatively, it measures some distance or similarity/dissimilarity between $p(x)$ and $q(x)$.

2.7.2 DL divergence

In information theory and machine learning, a very important concept is the Kullback–Leibler (KL) divergence, which is a distance measure between two probability distributions. The KL divergence $D_{KL}(p, q)$ is often denoted by $D_{KL}(p||q)$ to highlight that it represents the difference or distance between $p(x)$ given $q(x)$:

$$D_{KL}(p, q) = D_{KL}(p||q) = \int p(x) \log\Big(\frac{p(x)}{q(x)}\Big) dx, \tag{2.79}$$

and this integral becomes summation for discrete random variables.

From its definition we can show that

$$\begin{aligned} D_{KL}(p, q) &= \int p(x) \log[p(x)/q(x)] dx \\ &= \int p(x) \log[p(x)] dx - \int p(x) \log[q(x)] dx \\ &= H(p, q) - H(p), \end{aligned} \tag{2.80}$$

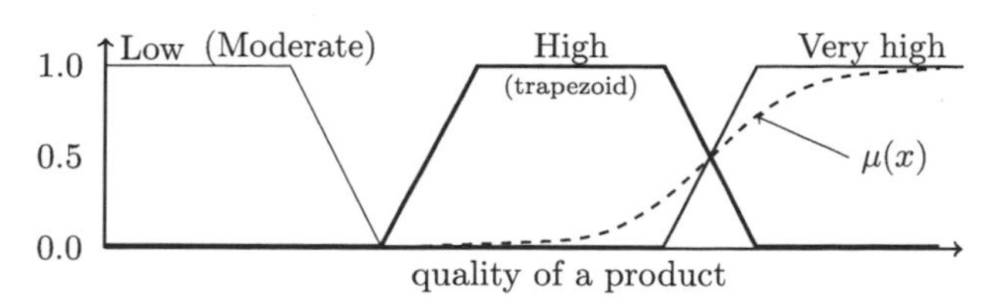

Figure 2.4 Membership functions for a fuzzy logic representation.

which is the difference between the cross entropy and the Shannon entropy [121]. Obviously, the KL divergence is zero when $p(x) = q(x)$.

2.8 Fuzzy rules

The mathematical foundations we have discussed so far are mainly based on the Boolean logics with clear and rigorous rules. In reality, many statements such as good quality of service and likeability may not be clear-cut as a simple binary yes–no answer, and in this case, we may have to deal with fuzzy rules or fuzzy logic in general. In some applications such as clustering and classification, it may be advantageous to use fuzzy rules. In fact, there is a whole fuzzy logic system [162]. For example, if we try to describe the quality of a product in terms of a set of particular measures, then we may say that the quality is low, high, very high, and so on. In this case, we can use a membership function μ, which is a piecewise linear function with the maximum value of 1.0 and the minimum value of 0.0. Such membership shows the degree of truth or the trueness of belonging to a description such as low or high (see Fig. 2.4). Often, the membership functions are of trapezoidal or triangular shape, though they may not be smooth. To have smooth membership functions, some special functions, such as the well-known sigmoid function

$$\mu(x) = \frac{1}{1+e^{-x}}, \tag{2.81}$$

are used to approximate certain membership as shown by the dashed curve in Fig. 2.4. In addition, a membership can be a Gaussian type of functions, or even an impulse-type Dirac delta functions.

In general, instead of the Boolean indicator function of binary values $\{0, 1\}$, the membership function can be a general nonnegative continuous function $\mu \in [0, 1]$ or even multivalued sets over $[0, 1]$. Furthermore, there is no requirement that it must be symmetric, and thus it can be any piecewise linear or piecewise smooth or nonsmooth functions.

Fuzzy sets and fuzzy systems have been used in control and decision-making [113] as well as fuzzy rule-based systems [122]. The interested readers can refer to more specialized literature [113].

2.9 Data mining and machine learning

Both data mining and machine learning are active areas of research, and they have a vast range of algorithms and technique. We use a unified approach in this book to relate algorithms in both areas to optimization algorithms.

2.9.1 Data mining

Data mining is a big area of data sciences, which aims to discover patterns and features in data, often large data sets. It includes regression, classification, clustering, detection of anomaly, and others. It also includes preprocessing, validation, summarization, and ultimately the making sense of the data sets.

The evolution of the Internet and the social media has resulted in the huge explosion of the data volumes and complexity, the so-called big data nowadays. Consequently, data mining has also expanded beyond the traditional data modeling such as regression and statistical models. For example, the aim of clustering is dividing n observations into some different clusters, based on certain clustering measures or objectives. Classification is dividing the data set into different classes with different labels such as normal or abnormal, yes or no.

There are a vast array of data mining methods. In this book, we will introduce the most widely used techniques.

2.9.2 Machine learning

Machine learning is an important area of artificial intelligence and computer science. Machine learning algorithms are a class of sophisticated algorithms such as supervised learning, unsupervised learning, semisupervised learning algorithms, and others. In general, there are a diverse range of algorithms in this category, including classification, linear regression, principal component analysis, logistic regression, decision trees, artificial neural networks, support vector machines, Bayesian networks, Boltzmann machine, deep belief networks, and others. It also includes optimization algorithms such as stochastic gradient descent and evolutionary algorithms.

Due to the diversity of such algorithms, there is a vast literature in machine learning and artificial intelligence. In this book, we introduce some of the most widely used algorithms with the emphasis on the core algorithms and their essential characteristics.

2.10 Notes on software

There are many different software packages, ranging from mathematics-oriented software such as Mathematica and Maple to statistical and simulation software such as R and WinBUGS.

- For symbolic computation and mathematics: Mathematica, Maple, and MuPAD (part of Matlab) are powerful. Also, free packages such as Axiom, Maxima, and SymPy are very versatile. A list of such packages can be found in https://en.wikipedia.org/wiki/List_of_computer_algebra_systems.
- For statistical modeling and computation, R, Matlab, Python, MiniTab, and SPSS all have powerful capabilities. For MCMC samplings, BUGS with Gibbs samplings is very powerful.[1]
- For data mining and machine learning, there are a vast range of software packages, from TensorFlow and Keras to R, Python, and RapidMiner. We will discuss these in more detail in late chapters. For other systems such as fuzzy systems, a review of software can be found in other literature such as [122].

[1] WinBUGS: http://www.mrc-bsu.cam.ac.uk/software/bugs.

Optimization algorithms

3

Contents

Optimization algorithms are diverse with many specialized techniques and a few general techniques. The algorithms for optimization include gradient-based algorithms, gradient-free algorithms, evolutionary algorithms and nature-inspired metaheuristics. We will mainly focus on the introduction of the gradient-based techniques due to their importance in data mining and machine learning. We will also briefly outline some gradient-free methods and metaheuristic algorithms. For a more detailed introduction, we refer the readers to the more advanced literature [22,5,7,159,161].

3.1 Gradient-based methods

Gradient-based methods are iterative methods that extensively use the gradient information of the objective function during iterations. Let us start with the simplest Newton method.

3.1.1 *Newton's method*

For minimization and maximization of a univariate function $f(x)$, it is equivalent to finding the roots of its gradient $g(x) = f'(x) = 0$. From Newton's root-finding

Introduction to Algorithms for Data Mining and Machine Learning. https://doi.org/10.1016/B978-0-12-817216-2.00010-7

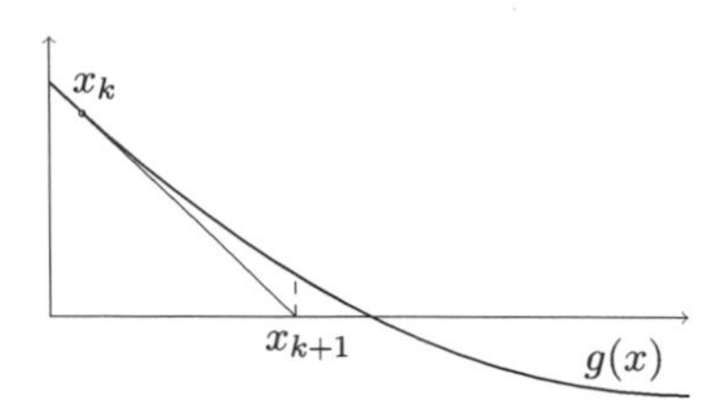

Figure 3.1 Newton's method and iterations.

algorithm (1.6) we have

$$x_{k+1} = x_k - \frac{g(x_k)}{g'(x_k)} = x_k - \frac{f'(x_k)}{f''(x_k)}, \tag{3.1}$$

where we have used $g'(x) = f''(x)$, implicitly assuming that these derivatives exist. The main idea of Newton's method is shown in Fig. 3.1.

Example 14

For a simple function $f(x) = (x-1)^2 = x^2 - 2x + 1$ in the real domain $x \in \mathbb{R}$, we know that its global minimum is $f_{\min} = 0$ at $x_* = 1$. Let us use Newton's formula (3.1) to find this solution starting from any value $x_0 = a > 0$. We have $f'(x) = 2x - 2$ and $f''(x) = 2$. From

$$x_{k+1} = x_k - \frac{f'(x_k)}{f''(x_k)} \tag{3.2}$$

we have

$$x_1 = x_0 - \frac{2x_0 - 2}{2} = a - \frac{2a-2}{2} = 1, \tag{3.3}$$

which gives the optimal solution in a single step. This shows that this algorithm is very efficient.

It is worth pointing out that this function is a special case because $f(x) = (x-1)^2$ is a convex function, and thus Newton's method can find the solution in a single step. In general, $f(x)$ is not convex with possible multiple solutions, and some care should be taken.

Let us revisit an earlier example (Example 3) where $f(x) = x^2 \exp(-x^2)$ has two maxima at $x_* = \pm 1$ and one minimum at $x_* = 0$. Then $f'(x) = 2x(1-x^2)e^{-x^2}$ and $f''(x) = 2e^{-x^2}(1 - 5x^2 + 2x^4)$. Thus, Newton's iterative formula becomes

$$\begin{aligned} x_{k+1} &= x_k - \frac{f'(x_k)}{f''(x_k)} = x_k - \frac{2x_k(1-x_k^2)e^{-x_k^2}}{2e^{-x_k^2}(1-5x_k^2+2x_k^4)} \\ &= x_k - \frac{x_k(1-x_k^2)}{1-5x_k^2+2x_k^4}, \end{aligned} \tag{3.4}$$

where we have used $\exp(-x_k^2) \neq 0$.

If we start with $x_0 = 0.8$, then we have

$$x_1 = 0.8 - \frac{0.8 \times (1 - 0.8^2)}{1 - 5 \times 0.8^2 + 2 \times 0.8^4} \approx 1.0085747 \tag{3.5}$$

and

$$x_2 = 0.999961, \quad x_3 = 0.9999999, \tag{3.6}$$

which are very close to $x_* = 1.0$.

If we use $x_0 = 0.5$, then we have

$$x_1 = 3.5, \quad x_2 \approx 3.66414799, \quad x_3 \approx 3.818812, \tag{3.7}$$

which gradually moves toward infinity. In this case, the iterative sequence becomes divergent. We can never reach the optimal solution $x_* = 1.0$. Similarly, if we use $x_0 = 2.0$, then it leads a similar divergent sequence.

If we start with $x_0 = 0.2$, then we will get $x_* = 0$ in a few steps. However, if we start with $x_0 = 0.4$, then we have

$$x_1 = -0.93757962, \quad x_2 \approx -0.9988808587, \quad x_3 \approx -0.9999993, \tag{3.8}$$

which rapidly converges toward $x_* = -1.0$.

This highlights an important issue here. Different starting points x_0 can lead to completely different final solutions. The solution sequences will largely depend on the initial solution x_0. It seems that we cannot predict easily which final solutions the iterative procedure may produce. Through detailed mathematical analysis of the iterative formula, it may be possible to figure out critical points for bifurcation and different branches. However, a high nonlinearity of the iteration formula makes it difficult to see which branch a particular initial point may lead to. Even if it is possible, it may not worth the effort because it cannot be generalized.

In fact, this issue of the dependence of the final solutions on the initial point is almost universal for many optimization algorithms, especially for those based on gradient information. The only exception is linear programming and convex optimization. Various efforts have been dedicated to solve this issue so as to design optimization algorithms that are less dependent on (ideally, independent of) initial configuration. We will come back to this issue again in later chapters and will provide, whenever possible, various remedies to this key issue.

3.1.2 Newton's method for multivariate functions

Newton's method works for univariate functions. Now let us extend it to solve optimization problems for multivariate functions. For the minimization of a function $f(\boldsymbol{x})$, $\boldsymbol{x} = (x_1, x_2, ..., x_n)$, the essence of this method is

$$\boldsymbol{x}^{(k+1)} = \boldsymbol{x}^{(k)} + \alpha g(\nabla f, \boldsymbol{x}^{(k)}), \tag{3.9}$$

where α is the step size, which can vary during iterations, and $g(\nabla f, \boldsymbol{x}^{(k)})$ is a function of the gradient ∇f and the current location $\boldsymbol{x}^{(k)}$. Different methods use different forms of $g(\nabla f, \boldsymbol{x}^{(k)})$.

We know that Newton's method is a popular iterative method for finding the zeros of a nonlinear univariate function of $f(x)$ on the interval $[a, b]$. It can be modified for solving optimization problems because it is equivalent to finding the zeros of the first derivative $f'(\boldsymbol{x})$ once the objective function $f(\boldsymbol{x})$ is given.

For a given continuously differentiable function $f(\boldsymbol{x})$, we have the Taylor expansion about a known point $\boldsymbol{x} = \boldsymbol{x}_k$ (with $\Delta \boldsymbol{x} = \boldsymbol{x} - \boldsymbol{x}_k$)

$$f(\boldsymbol{x}) = f(\boldsymbol{x}_k) + (\nabla f(\boldsymbol{x}_k))^T \Delta \boldsymbol{x} + \frac{1}{2} \Delta \boldsymbol{x}^T \nabla^2 f(\boldsymbol{x}_k) \Delta \boldsymbol{x} + \cdots,$$

which is minimized near a critical point when $\Delta \boldsymbol{x}$ is the solution of the linear equation

$$\nabla f(\boldsymbol{x}_k) + \nabla^2 f(\boldsymbol{x}_k) \Delta \boldsymbol{x} = 0, \quad \text{or} \quad \boldsymbol{x} = \boldsymbol{x}_k - \boldsymbol{H}^{-1} \nabla f(\boldsymbol{x}_k), \tag{3.10}$$

where $\boldsymbol{H} = \nabla^2 f(\boldsymbol{x}_k)$ is the Hessian matrix. If the iteration procedure starts from the initial vector $\boldsymbol{x}^{(0)}$ (usually taken to be a guessed point in the domain), then Newton's iteration formula for the kth iteration is

$$\boldsymbol{x}^{(k+1)} = \boldsymbol{x}^{(k)} - \boldsymbol{H}^{-1}(\boldsymbol{x}^{(k)}) \nabla f(\boldsymbol{x}^{(k)}). \tag{3.11}$$

It is worth pointing out that if $f(\boldsymbol{x})$ is quadratic, then the solution can be found exactly in a single step. However, this method may become tricky for nonquadratic functions, especially when we have to calculate a large Hessian matrix.

It can usually be time-consuming to calculate the Hessian matrix for second derivatives. A good alternative is to use an identity matrix to approximate the Hessian by using $\boldsymbol{H}^{-1} = \boldsymbol{I}$, and we have the quasi-Newton method

$$\boldsymbol{x}^{(k+1)} = \boldsymbol{x}^{(k)} - \alpha \boldsymbol{I} \, \nabla f(\boldsymbol{x}^{(k)}), \tag{3.12}$$

where $\alpha \in (0, 1)$ is a step size. In this case, the method is essentially the steepest descent method.

Though gradient-based methods can be very efficient, the final solution tends to depend on the starting point. If the starting point is very far away from the optimal solution, the algorithm can either reach a completely different solution for multimodal problems or simply fail in some cases. Therefore there is no guarantee that the global optimal solution can be found.

It is worth pointing out that there are many variations of the steepest descent methods. If such optimization aims is to find the maximum, then this method becomes a *hill-climbing* method because the aim is to climb up the hill to the highest peak.

3.1.3 Line search

In the steepest descent method, there are two important parts, the descent direction and the step size (or how far to descend). The calculations of the exact step size may be very time consuming. In reality, we intend to find the right descent direction. Then a reasonable amount of descent, not necessarily the exact amount, during each iteration will usually be sufficient. For this, we essentially use a line search method.

To find the local minimum of the objective function $f(\boldsymbol{x})$, we try to search along a descent direction $\boldsymbol{s}_k$ with an adjustable step size α_k so that

$$\psi(\alpha_k) = f(\boldsymbol{x}_k + \alpha_k \boldsymbol{s}_k) \tag{3.13}$$

decreases as much as possible, depending on the value of α_k. Loosely speaking, a reasonably right step size should satisfy the Wolfe conditions

$$f(\boldsymbol{x}_k + \alpha_k \boldsymbol{s}_k) \leq f(\boldsymbol{x}_k) + \gamma_1 \alpha_k \boldsymbol{s}_k^T \nabla f(\boldsymbol{x}_k) \tag{3.14}$$

and

$$\boldsymbol{s}_k^T \nabla f(\boldsymbol{x}_k + \alpha_k \boldsymbol{s}_k) \geq \gamma_2 \boldsymbol{s}_k^T \nabla f(\boldsymbol{x}_k), \tag{3.15}$$

where $0 < \gamma_1 < \gamma_2 < 1$ are algorithm-dependent parameters. The first condition is a sufficient decrease condition for α_k, often called the Armijo condition or rule, whereas the second inequality is often referred to as the curvature condition. For most functions, we can use $\gamma_1 = 10^{-4}$ to 10^{-2} and $\gamma_2 = 0.1$ to 0.9. These conditions are usually sufficient to ensure the algorithm converge in most cases; however, stronger conditions may be needed for some tough functions.

The basic steps of the line search method can be summarized in Algorithm 2.

Algorithm 2 Line search method.

1: Initial guess $\boldsymbol{x}_0$ at $k = 0$
2: **while** $\|\nabla f(\boldsymbol{x}_k)\| >$ accuracy **do**
3: Find the search direction $\boldsymbol{s}_k = -\nabla f(\boldsymbol{x}_k)$
4: Solve for α_k by decreasing $f(\boldsymbol{x}_k + \alpha \boldsymbol{s}_k)$ significantly
5: satisfying the Wolfe conditions
6: Update the result $\boldsymbol{x}_{k+1} = \boldsymbol{x}_k + \alpha_k \boldsymbol{s}_k$
7: $k \leftarrow k + 1$
8: **end while**

3.2 Variants of gradient-based methods

Over many decades, a class of gradient-based methods have been developed. They all use some forms of gradient information, though their algorithmic procedures can be very different. Here, we introduce a few commonly used variants. In the context of machine learning, there are some comprehensive reviews. For example, Ruder [125] provided an overview of gradient descent optimization algorithms.

3.2.1 Stochastic gradient descent

In many optimization problems, especially in deep learning, the objective function or risk function to be minimized can be written in the following form:

$$E(\boldsymbol{w}) = \frac{1}{m}\sum_{i=1}^{m} f_i(\boldsymbol{x}_i, \boldsymbol{w}) = \frac{1}{m}\sum_{i=1}^{m}\big[u_i(\boldsymbol{x}_i, \boldsymbol{w}) - \bar{y}_i\big]^2, \tag{3.16}$$

where

$$f_i(\boldsymbol{x}_i, \boldsymbol{w}) = \big[u_i(\boldsymbol{x}_i, \boldsymbol{w}) - \bar{y}_i\big]^2. \tag{3.17}$$

Here, $\boldsymbol{w} = (w_1, w_2, \ldots, w_K)^T$ is a parameter vector such as the weights in a neural network. In addition, $\bar{y}_i$ $(i = 1, 2, \ldots, m)$ are the target or real data (data points or data sets), whereas $u_i(\boldsymbol{x}_i)$ are the predicted values based on the inputs $\boldsymbol{x}_i$ by a model such as the models based on trained neural networks.

The standard gradient descent for finding new weight parameters in terms of iterative formula can be written as

$$\boldsymbol{w}^{t+1} = \boldsymbol{w}^t - \frac{\eta}{m}\sum_{i=1}^{m}\nabla f_i, \tag{3.18}$$

where $0 < \eta \leq 1$ is the learning rate or step size. Here, the gradient ∇f_i is with respect to $\boldsymbol{w}$. This requires the calculations of m gradients. When m is large and the number of iteration t is large, this can be very expensive.

To save computation, the true gradient can be approximated by the gradient at a single value at f_i instead of all m values, that is,

$$\boldsymbol{w}^{t+1} = \boldsymbol{w}^t - \eta_t \nabla f_i, \tag{3.19}$$

where η_t is the learning rate at iteration t, which may vary with iterations. Though this is a crude estimate at a randomly selected point i at iteration t to the true gradient, the computation costs have dramatically reduced by a factor of $1/m$. Due to the random nature of the selection of a sample i (which can be very different at each iteration), this way of calculating gradient is called stochastic gradient. The method based on such crude estimation of gradients is called stochastic gradient descent (SGD) for minimization or stochastic gradient ascent (SGA) for maximization.

The learning rate η_t should be reduced gradually. For example, a commonly used reduction of learning rates is

$$\eta_t = \frac{1}{1 + \beta t}, \quad t = 1, 2, \ldots, \tag{3.20}$$

where $\beta > 0$ is a hyper-parameter (see Fig. 3.2).

Bottou showed that SGD almost surely converges if

$$\sum_t \eta_t = \infty, \quad \sum_t \eta_t^2 < \infty. \tag{3.21}$$

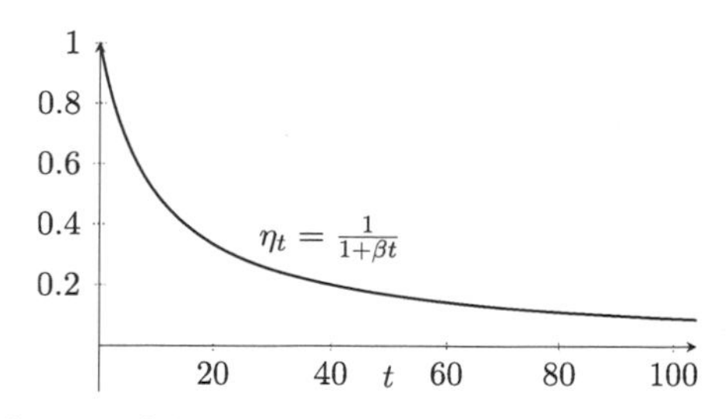

Figure 3.2 Monotonic decrease of the learning rate.

The best convergence rate is $\eta_t \sim 1/t$ with the averaged residual error decreasing as $E \sim 1/t$ [18].

It is worth pointing out the stochastic gradient descent is not the direct descent in the true gradient sense, but the descent is in terms of average or expectation. Thus, the paths can still be zig-zag, sometimes, it may be up the gradient, not necessarily all the way down the gradient directions, but the overall computation efficiency is usually much higher than the true gradient descent for large-scale problems. Therefore, it is widely used for deep learning problems and large-scale problems.

3.2.2 Subgradient method

All the gradient-based methods mentioned assume implicitly that the functions are differentiable. In the case of nondifferentiable functions, we have to use the subgradient method for non-differential convex functions or more generally gradient-free methods for nonlinear functions to be introduced later in this chapter.

For nondifferentiable convex functions, the subgradient vectors $\boldsymbol{v}_k$ can be defined by

$$f(\boldsymbol{x}) - f(\boldsymbol{x}_k) \geq \boldsymbol{v}_k^T(\boldsymbol{x} - \boldsymbol{x}_k), \tag{3.22}$$

and Newton's iteration formula can be replaced by

$$\boldsymbol{x}^{k+1} = \boldsymbol{x}^k - \alpha_k \boldsymbol{v}_k, \tag{3.23}$$

where α_k is the step size at iteration k. As the iteration formula involves the subgradient $\boldsymbol{v}_k = \partial f(\boldsymbol{x}_k)$ calculated at iteration k, the method is called the subgradient method.

It is worth pointing out that since there are many arbitrary subgradients (see Fig. 3.3), the subgradient calculated at $\boldsymbol{x}_k$ may not be in the desirable direction. Some choices such as choosing greater values of the norm $\boldsymbol{v}$ can be expected.

In addition, though a constant step size $\alpha_k = \alpha$ where $0 < \alpha < 1$ can work well in many cases, it is desirable that the step size α_k should vary and be scaled when appropriate. For example, a commonly used scheme for varying step sizes is $\alpha_k \geq 0$, $\sum_{k=1}^{\infty} \alpha_k^2 < \infty$, and $\lim_{k\to\infty} \alpha_k = 0$.

The subgradient method is still used in practice, and it can be very effective in combination with the stochastic gradient method, which leads to a class of so-called

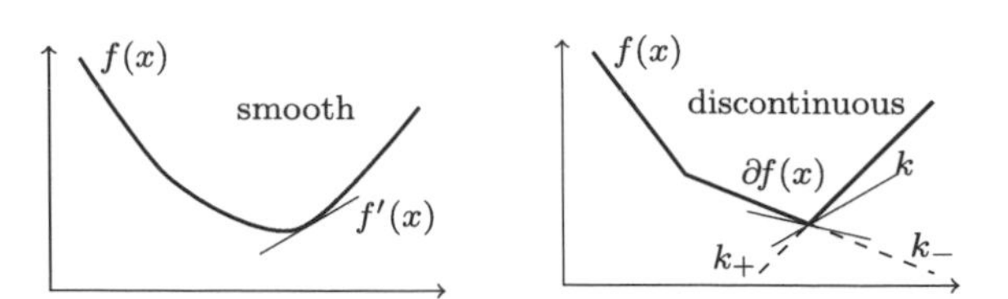

Figure 3.3 Gradient of smooth $f(x)$ (left) and subgradients (right) of $f(x)$ with discontinuity.

stochastic subgradient methods. The convergence can be proved, and the interested readers can refer to more advanced literature such as Bertsbekas et al. [16].

The limitation of the subgradient method is that it is mainly for convex functions. In case of general nonlinear nondifferentiable nonconvex functions, we can use gradient-free methods, and we will introduce some of these methods in the next section.

3.2.3 Conjugate gradient method

The method of conjugate gradient belongs to a wider class of the so-called Krylov subspace iteration methods. The conjugate gradient method was pioneered by Magnus Hestenes, Eduard Stiefel, and Cornelius Lanczos in the 1950s. It was named as one of the top 10 algorithms of the 20th century.

The conjugate gradient method can be used to solve the linear system

$$\boldsymbol{A}\boldsymbol{u} = \boldsymbol{b}, \tag{3.24}$$

where $\boldsymbol{A}$ is often a symmetric positive definite matrix. The above system is equivalent to minimizing the function

$$f(\boldsymbol{u}) = \frac{1}{2}\boldsymbol{u}^T \boldsymbol{A}\boldsymbol{u} - \boldsymbol{b}^T \boldsymbol{u} + c, \tag{3.25}$$

where c is a constant and can be taken to be zero. We can easily see that $\nabla f(\boldsymbol{u}) = 0$ leads to $\boldsymbol{A}\boldsymbol{u} = \boldsymbol{b}$.

In general, the size of $\boldsymbol{A}$ can be very large and sparse with $n > 100{,}000$, but it is not required that $\boldsymbol{A}$ is strictly symmetric positive definite. In fact, the main condition is that $\boldsymbol{A}$ should be a normal matrix. A square matrix $\boldsymbol{A}$ is called normal if $\boldsymbol{A}^T\boldsymbol{A} = \boldsymbol{A}\boldsymbol{A}^T$. Therefore a symmetric matrix is a normal matrix, so is an orthogonal matrix because an orthogonal matrix $\boldsymbol{Q}$ satisfies $\boldsymbol{Q}\boldsymbol{Q}^T = \boldsymbol{Q}^T\boldsymbol{Q} = \boldsymbol{I}$.

The theory behind these iterative methods is closely related to the Krylov subspace $\mathcal{K}_k$ spanned by $\boldsymbol{A}$ and $\boldsymbol{b}$ and defined by

$$\mathcal{K}_k(\boldsymbol{A}, \boldsymbol{b}) = \{\boldsymbol{I}\boldsymbol{b}, \boldsymbol{A}\boldsymbol{b}, \boldsymbol{A}^2\boldsymbol{b}, \ldots, \boldsymbol{A}^{n-1}\boldsymbol{b}\}, \tag{3.26}$$

where $\boldsymbol{A}^0 = \boldsymbol{I}$.

If we use an iterative procedure to obtain the approximate solution $\boldsymbol{u}_k$ to $\boldsymbol{A}\boldsymbol{u} = \boldsymbol{b}$ at the kth iteration, then the residual is given by

$$\boldsymbol{r}_k = \boldsymbol{b} - \boldsymbol{A}\boldsymbol{u}_k, \tag{3.27}$$

which is essentially the negative gradient $\nabla f(\boldsymbol{u}_k)$. The search direction vector in the conjugate gradient method is subsequently determined by

$$\boldsymbol{d}_{k+1} = \boldsymbol{r}_k - \frac{\boldsymbol{d}_k^T \boldsymbol{A} \boldsymbol{r}_k}{\boldsymbol{d}_k^T \boldsymbol{A} \boldsymbol{d}_k} \boldsymbol{d}_k. \tag{3.28}$$

The solution often starts with an initial guess $\boldsymbol{u}_0$ at $k = 0$ and proceeds iteratively. The above steps can compactly be written as

$$\boldsymbol{u}_{k+1} = \boldsymbol{u}_k + \alpha_k \boldsymbol{d}_k, \quad \boldsymbol{r}_{k+1} = \boldsymbol{r}_k - \alpha_k \boldsymbol{A} \boldsymbol{d}_k, \tag{3.29}$$

and

$$\boldsymbol{d}_{k+1} = \boldsymbol{r}_{k+1} + \beta_k \boldsymbol{d}_k, \tag{3.30}$$

where

$$\alpha_k = \frac{\boldsymbol{r}_k^T \boldsymbol{r}_k}{\boldsymbol{d}_k^T \boldsymbol{A} \boldsymbol{d}_k}, \qquad \beta_k = \frac{\boldsymbol{r}_{k+1}^T \boldsymbol{r}_{k+1}}{\boldsymbol{r}_k^T \boldsymbol{r}_k}. \tag{3.31}$$

Iterations stop when a prescribed accuracy is reached. This can easily be programmed in any programming language, especially Matlab and Python.

It is worth pointing out that the initial guess $\boldsymbol{r}_0$ can be any educated guess; however, $\boldsymbol{d}_0$ should be taken as $\boldsymbol{d}_0 = \boldsymbol{r}_0$, since otherwise the algorithm may not converge.

3.3 Optimizers in deep learning

Gradient-based optimizers are widely used in machine learning, especially in many recent studies in deep learning. For a recent review on such optimizers, we refer the readers to Bengio et al. [12] and Ruder [125].

Stochastic gradient descent (SGD) methods can be very efficient if used properly. However, such methods can have strong oscillations for objectives with narrow valleys. A useful modification is introducing a momentum term so as to damp such potential oscillations [116]. Usually, a gradient-based method for minimizing the objective $f(\boldsymbol{x})$ is using the iterative formula

$$\boldsymbol{x}^{(k+1)} = \boldsymbol{x}^{(k)} - \eta \nabla f(\boldsymbol{x}^{(k)}), \tag{3.32}$$

where $0 < \eta < 1$ is the learning rate.

- In momentum-based methods, the main additional step is to use

$$\boldsymbol{u}^{(k+1)} = \gamma \boldsymbol{u}^{(k)} + \eta \nabla f(\boldsymbol{x}^{(k)}), \quad 0 < \gamma < 1, \tag{3.33}$$

  and then update the increment by

$$\boldsymbol{x}^{(k+1)} = \boldsymbol{x}^{(k)} - \boldsymbol{u}^{(k+1)}. \tag{3.34}$$

This essentially uses momentum to speed up the downhill moves, leading to a potentially higher rate of convergence. Typically, the value $\gamma = 0.9$ is used [116,125].

- Another related speedup is the Nesterov accelerated gradient (NAG) method [110], which uses a predicted or modified step

$$\boldsymbol{u}^{(k+1)} = \gamma \boldsymbol{u}^{(k)} + \eta \nabla f\big(\boldsymbol{x}^{(k)} - \gamma \boldsymbol{u}^{(k)}\big), \tag{3.35}$$

$$\boldsymbol{x}^{(k+1)} = \boldsymbol{x}^{(k)} - \boldsymbol{u}^{(k+1)}. \tag{3.36}$$

- The adaptive subgradient method AdaGrad uses an adaptive learning rate for every variable x_i [41], that is, η is replaced by

$$\eta \leftarrow \frac{\eta}{\sqrt{G_k + \epsilon}}, \tag{3.37}$$

where $\epsilon > 0$ is a small number to avoid division by zero; $\epsilon = 10^{-8}$ is usually used. $G_k = [g_{ii}] \in \mathbb{R}^{n \times n}$ is a diagonal matrix where the ith diagonal element is the sum of the squares of the gradient with respect to x_i $(i = 1, 2, \ldots, n)$ up to iteration k. Thus, the iterative equation becomes

$$\boldsymbol{x}^{(k+1)} = \boldsymbol{x}^{(k)} - \frac{\eta}{\sqrt{G_k + \epsilon}} \otimes \nabla f(\boldsymbol{x}^{(k)}), \tag{3.38}$$

where $\otimes$ means the elementwise product so that this formula works for each variable x_i.

- One of the disadvantage of the AdaGrad method is that it has to somehow store the gradient values

$$g_k = \nabla f(\boldsymbol{x}^{(k)}) \tag{3.39}$$

so as to calculate G_k. One improvement is to use the running average $E_k(g^2)$ of squared gradients at iteration k, which means that the update simply becomes

$$E_k(g_k^2) = \gamma E_k(g_{k-1}^2) + (1-\gamma) g_k^2. \tag{3.40}$$

The parameter increment is given by

$$\Delta \boldsymbol{x}_k = -\frac{\eta}{\mathrm{RMS}[g_k]} g_k, \tag{3.41}$$

where $\mathrm{RMS}[g_k] = \sqrt{E[g_k^2] + \epsilon}$ is the root mean squared (RMS) error. The exponential decay of the average for the increment is calculated by

$$E[(\Delta x_k)^2] = \gamma E[(\Delta x_{k-1})^2] + (1-\gamma)(\Delta \boldsymbol{x}_k)^2. \tag{3.42}$$

Thus the iterative formula becomes

$$\boldsymbol{x}^{(k+1)} = \boldsymbol{x}^{(k)} - \frac{\sqrt{E[(\Delta \boldsymbol{x}_{k-1})^2] + \epsilon}}{\mathrm{RMS}[g_k]} g_k, \tag{3.43}$$

which leads to the AdaDelta method [164].

- Another optimizer is RMSprop, which was introduced independently by Geofrey Hinton in his course on neural networks for machine learning, detailed in his unpublished lecture notes.[1] The main idea is to use a running average of its recent gradient magnitude with an exponential decay. RMSprop has some similarity to AdaDelta method, and the running average becomes the same when $\gamma = 0.9$ in Eq. (3.40). The weight update becomes

$$\boldsymbol{x}^{(k+1)} = \boldsymbol{x}^{(k)} - \frac{\eta}{\sqrt{E_k(g_k^2) + \epsilon}} g_k. \tag{3.44}$$

  The learning rate is usually set to $\eta = 0.001$.
- The very popular Adam optimizer was developed in 2014 by Kingma and Ba [88]. The influence of the previously stored gradient is decaying exponentially. For the Adam optimizer, the main steps use both the first moment m_1 and the second moment m_2 at each iteration k, which correspond to the mean and the uncentered variance. We have

$$\begin{cases} m_1^{(k)} = \alpha m_1^{(k-1)} + (1-\alpha) g_k, \\ m_2^{(k)} = \beta m_2^{(k-1)} + (1-\beta) g_k^2, \end{cases} \tag{3.45}$$

  where α, β are parameters. However, both moment estimates are somehow biased, and thus corrections are needed [88]. The iterative formula becomes

$$\boldsymbol{x}^{(k+1)} = \boldsymbol{x}^{(k)} - \frac{\eta}{\sqrt{\bar{m}_2^{(k)}} + \epsilon} \bar{m}_1^{(k)}, \tag{3.46}$$

  where

$$\bar{m}_1^{(k)} = \frac{m_1^{(k)}}{1-\alpha^k}, \quad \bar{m}_2^{(k)} = \frac{m_2^{(k)}}{1-\beta^k}. \tag{3.47}$$

  Here α^k and β^k are the values to the power k. The values of the parameters $\alpha = 0.9$, $\beta = 0.999$, $\eta = 0.001$, and $\epsilon = 10^{-8}$ can be used [88]. The initial values $m_1^{(0)} = m_2^{(0)} = 0$ can be used.

There are many other variants of gradient descent methods such as AdaMax, Nadam, and others. We refer the interested readers to more specialized literature such as Ruder [125], Bengio [10], and Bengio et al. [12].

All the algorithms mentioned have been implemented in major machine learning packages such as TensorFlow, Python, and others.

[1] http://www.cs.toronto.edu/~tijmen/csc321/slides/lecture_slides_lec6.pdf.

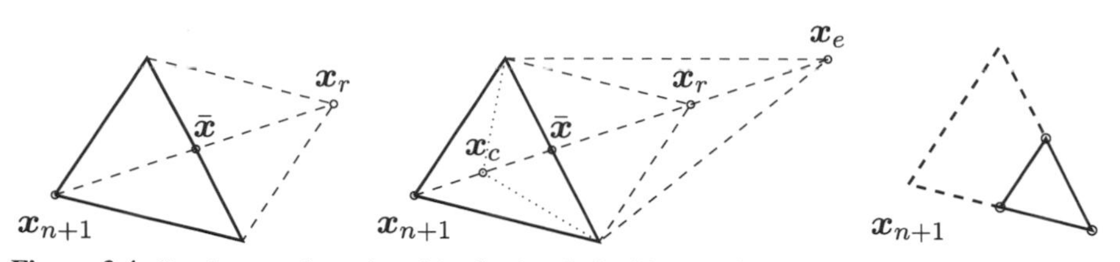

Figure 3.4 Simplex transformation: (a) reflection (left), (b) expansion and contraction (middle), (c) reduction (right).

3.4 Gradient-free methods

Though gradient-based methods are very efficient, they need to calculate derivatives during iterations. For some problems, the computation of derivatives can be expensive. For problems with discontinuous objectives, it is not possible to calculate such derivatives. In this case, methods that do not require derivatives are preferred. Such derivative-free or gradient-free methods can also be very effective.

The Nelder–Mead method is a downhill simplex algorithm for unconstrained optimization without using derivatives, and it was first developed in 1965 by Nelder and Mead [109]. This is one of widely used traditional methods since its computational effort is relatively small and is something to get a quick grasp of the optimization problem. The basic idea of this method is to use the flexibility of the constructed simplex via amoeba-style manipulations by reflection, expansion, contraction, and reduction (see Fig. 3.4). In some books, such as the best-known *Numerical Recipes*, it is also called the "Amoeba algorithm" [115]. It is worth pointing out that this downhill simplex method has nothing to do with the simplex method for linear programming.

In the n-dimensional space, a simplex, which is a generalization of a triangle on a plane, is a convex hull with $n + 1$ distinct points. For simplicity, a simplex in the n-dimensional space is referred to as an n-simplex. Therefore a 1-simplex is a line segment, a 2-simplex is a triangle, a 3-simplex is a tetrahedron, and so on.

There are a few variants of the algorithm that use slightly different ways of constructing initial simplex and convergence criteria. However, the fundamental procedure is the same (see Algorithm 3).

The first step is constructing an initial n-simplex with $n + 1$ vertices and evaluating the objective function at the vertices. Then, by ranking the objective values and reordering the vertices, we have an ordered set, so that

$$f(\boldsymbol{x}_1) \leq f(\boldsymbol{x}_2) \leq \cdots \leq f(\boldsymbol{x}_{n+1}) \tag{3.48}$$

at $\boldsymbol{x}_1, \boldsymbol{x}_2, \ldots, \boldsymbol{x}_{n+1}$, respectively. As the downhill simplex method is for minimization, we use the convention that $\boldsymbol{x}_{n+1}$ is the worse point (solution) and $\boldsymbol{x}_1$ is the best solution. Then, at each iteration, similar ranking manipulations are carried out.

Algorithm 3 Nelder–Mead (downhill simplex) method.

1: Initialize a simplex with $n+1$ vertices in n dimension.
2: **while** (stop criterion is not true) **do**
3: Reorder the points so that $f(\boldsymbol{x}_1) \leq f(\boldsymbol{x}_2) \leq \cdots \leq f(\boldsymbol{x}_{n+1})$
with $\boldsymbol{x}_1$ being the best and $\boldsymbol{x}_{n+1}$ being the worse (highest value)
4: Find the centroid $\bar{\boldsymbol{x}}$ using $\bar{\boldsymbol{x}} = \sum_{i=1}^{n} \boldsymbol{x}_i/n$ excluding $\boldsymbol{x}_{n+1}$.
5: Generate a trial point via the reflection of the worse vertex
6: $\boldsymbol{x}_r = \bar{\boldsymbol{x}} + \alpha(\bar{\boldsymbol{x}} - \boldsymbol{x}_{n+1})$ where $\alpha > 0$ (typically $\alpha = 1$)
7: **if** $f(\boldsymbol{x}_1) \leq f(\boldsymbol{x}_r) < f(\boldsymbol{x}_n)$ **then**
8: $\boldsymbol{x}_{n+1} \leftarrow \boldsymbol{x}_r$;
9: **end if**
10: **if** $f(\boldsymbol{x}_r) < f(\boldsymbol{x}_1)$ **then**
11: Expand in the direction of reflection $\boldsymbol{x}_e = \boldsymbol{x}_r + \beta(\boldsymbol{x}_r - \bar{\boldsymbol{x}})$
12: **if** $(f(\boldsymbol{x}_e) < f(\boldsymbol{x}_r))$ $\boldsymbol{x}_{n+1} \leftarrow \boldsymbol{x}_e$ **else** $\boldsymbol{x}_{n+1} \leftarrow \boldsymbol{x}_r$; **end**
13: **end if**
14: **if** $f(\boldsymbol{x}_r) > f(\boldsymbol{x}_n)$ **then**
15: Contract by $\boldsymbol{x}_c = \boldsymbol{x}_{n+1} + \gamma(\bar{\boldsymbol{x}} - \boldsymbol{x}_{n+1})$;
16: **if** $f(\boldsymbol{x}_c) < f(\boldsymbol{x}_{n+1})$ **then** $\boldsymbol{x}_{n+1} \leftarrow \boldsymbol{x}_c$;
17: **else** Reduction $\boldsymbol{x}_i = \boldsymbol{x}_1 + \delta(\boldsymbol{x}_i - \boldsymbol{x}_1)$, $(i = 2, \ldots, n+1)$; **end**
18: **end if**
19: **end while**

Then, we have to calculate the centroid $\boldsymbol{x}$ of the current simplex excluding the worst vertex $\boldsymbol{x}_{n+1}$:

$$\bar{\boldsymbol{x}} = \frac{1}{n} \sum_{i=1}^{n} \boldsymbol{x}_i. \tag{3.49}$$

Using the centroid as the basis point, we try to find the reflection of the worse point $\boldsymbol{x}_{n+1}$ by

$$\boldsymbol{x}_r = \bar{\boldsymbol{x}} + \alpha(\bar{x} - \boldsymbol{x}_{n+1}) \qquad (\alpha > 0), \tag{3.50}$$

though the typical value of $\alpha = 1$ is often used.

Whether the new trial solution is accepted or not and how to update the new vertex depends on the objective function at $\boldsymbol{x}_r$. There are three possibilities:

- If $f(\boldsymbol{x}_1) \leq f(\boldsymbol{x}_r) < f(\boldsymbol{x}_n)$, then replace the worst vertex $\boldsymbol{x}_{n+1}$ by $\boldsymbol{x}_r$, that is, $\boldsymbol{x}_{n+1} \leftarrow \boldsymbol{x}_r$.
- If $f(\boldsymbol{x}_r) < f(\boldsymbol{x}_1)$, which means the objective improves, then we seek a more bold move to see if we can improve the objective even further by moving or expanding the vertex further along the line of reflection to a new trial solution

$$\boldsymbol{x}_e = \boldsymbol{x}_r + \beta(\boldsymbol{x}_r - \bar{\boldsymbol{x}}), \tag{3.51}$$

where $\beta = 2$. Now we have to test if $f(\boldsymbol{x}_e)$ improves even better. If $f(\boldsymbol{x}_e) < f(\boldsymbol{x}_r)$, then we accept it and update $\boldsymbol{x}_{n+1} \leftarrow \boldsymbol{x}_e$; otherwise, we can use the result of the reflection, that is, $\boldsymbol{x}_{n+1} \leftarrow \boldsymbol{x}_r$.

- If there is no improvement or $f(\boldsymbol{x}_r) > f(\boldsymbol{x}_n)$, then we have to reduce the size of the simplex while maintaining the best sides. This is the contraction

$$\boldsymbol{x}_c = \boldsymbol{x}_{n+1} + \gamma(\bar{\boldsymbol{x}} - \boldsymbol{x}_{n+1}), \tag{3.52}$$

where $0 < \gamma < 1$, though $\gamma = 1/2$ is usually used. If $f(\boldsymbol{x}_c) < f(\boldsymbol{x}_{n+1})$, then we update $\boldsymbol{x}_{n+1} \leftarrow \boldsymbol{x}_c$.

If all these steps fail, then we have to reduce the size of the simplex toward the best vertex $\boldsymbol{x}_1$. This is the reduction step

$$\boldsymbol{x}_i = \boldsymbol{x}_1 + \delta(\boldsymbol{x}_i - \boldsymbol{x}_1) \qquad (i = 2, 3, \ldots, n+1). \tag{3.53}$$

Then, we go to the first step of the iteration process and start over again.

There are many other gradient-free optimization algorithms, and swarm intelligence-based algorithms are mostly gradient-free.

3.5 Evolutionary algorithms and swarm intelligence

The literature on evolutionary algorithms and nature-inspired algorithms is expanding rapidly. Most of these algorithms have drawn inspiration from evolutionary characteristics of biological or natural systems, and these algorithms form the majority of the evolutionary algorithms. In addition, recent algorithms use multiple agents to mimic the collective or social characteristics of swarming systems, and these algorithms somehow can simulate certain aspects of swarm intelligence (SI). In general, a vast majority of these algorithms are nature-inspired algorithms.

There are a large number of different algorithms, including evolutionary strategy, simulated annealing (SA), colony optimization (ACO), bees algorithms, genetic algorithm, bat algorithm, cuckoo search, differential evolution, firefly algorithm, particle swarm optimization, flower pollination algorithm, harmony search, memetic algorithm, and others. To introduce these algorithms systematically, an entire book [159] is required; therefore, we will only briefly introduce some of the most recent and widely used nature-inspired optimization algorithms [159,158].

Though many algorithms such as ACO, SA, and others belong to this category, we will not introduce them since they are not widely used yet in machine learning and data mining. We start with genetic algorithms.

3.5.1 Genetic algorithm

The genetic algorithm (GA), developed by John Holland and his collaborators in the 1960s and 1970s, is a model or abstraction of biological evolution based on Charles

Darwin's theory of natural selection. The genetic algorithm (GA) is an evolutionary algorithm and probably the most widely used. It is becoming a conventional and classic method. However, it does have fundamental genetic operators that have inspired many later algorithms, so we will introduce it in detail. There are many variants of the genetic algorithm, and they now form a class of genetic algorithms [75,57]. The essence of genetic algorithms involves the encoding of an objective function as arrays of bits or character strings to represent the chromosomes, the manipulation operations of strings by genetic operators, and the selection according to their fitness with the aim of finding a solution to the problem concerned. This is often done by the following procedure: 1) encoding of solutions into strings; 2) defining a fitness function and selection criterion; 3) creating a population of individuals and evaluating their fitness; 4) evolving the population by generating new solutions using crossover, mutation, fitness-proportionate reproduction; 5) selecting new solutions according to their fitness and replacing the old population by better individuals; and 6) decoding the results to obtain the solution(s) to the problem.

An important issue is the formulation or choice of an appropriate fitness function that determines the selection criterion in a particular problem. For the minimization of $f(x)$ using genetic algorithms, one simple way of constructing a fitness function is to use the simplest form $F(\boldsymbol{x}) = A - f(\boldsymbol{x})$ with A being a large constant (though $A = 0$ will do), and thus the objective is maximizing the fitness function. However, there are many different ways of defining a fitness function. For example, we can use the individual fitness assignment relative to the whole population

$$F(\boldsymbol{x}_i) = \frac{f(\boldsymbol{x}_i))}{\sum_{i=1}^{N} f(\boldsymbol{x}_i)}, \tag{3.54}$$

where N is the population size. The appropriate form of the fitness function will ensure that the solutions with higher fitness will be selected efficiently. Poorly defined fitness functions may result in incorrect or meaningless solutions.

Another important issue is the choice of various parameters. The crossover probability p_c is usually very high, typically in the range 0.7–0.99. On the other hand, the mutation probability p_m is usually small (usually 0.001–0.05). If p_c is too small, then the crossover occurs sparsely, which is not efficient for evolution. If the mutation probability is too high, then the diversity of the population may be too high, which makes it harder for the system to converge.

The selection criterion is also important so as to select the current population so that the best individuals with higher fitness are preserved and passed on to the next generation, which is often carried out in association with a certain elitism. The basic elitism is to select the most fit individual (in each generation), which will be carried over to the new generation without being modified by genetic operators. This ensures that the best solution is achieved more quickly.

3.5.2 Differential evolution

Differential evolution (DE) was developed in 1997 by Storn and Price [137]. It is a vector-based algorithm, which has some similarity to pattern search and genetic algorithms due to its use of crossover and mutation. DE is a stochastic search algorithm with self-organizing tendency and does not use the information of derivatives. Thus it is a population-based derivative-free method. In addition, DE uses real numbers as solution strings, and thus no encoding and decoding are needed.

For a D-dimensional optimization problem with D parameters, a population of n solution vectors is initially generated, and we have $\boldsymbol{x}_i$ for $i = 1, 2, \ldots, n$. For each solution $\boldsymbol{x}_i$ at any generation t, we use the conventional notation

$$\boldsymbol{x}_i^t = (x_{1,i}^t, x_{2,i}^t, \ldots, x_{D,i}^t), \tag{3.55}$$

which consists of D components in the D-dimensional space. This vector can be considered as the chromosomes or genomes.

Differential evolution consists of three main steps: mutation, crossover, and selection.

Mutation is carried out by the mutation scheme. For each vector $\boldsymbol{x}_i$ at any time or generation t, we first randomly choose three distinct vectors $\boldsymbol{x}_p$, $\boldsymbol{x}_q$, and $\boldsymbol{x}_r$ at t and then generate a so-called donor vector by the mutation scheme

$$\boldsymbol{v}_i^{t+1} = \boldsymbol{x}_p^t + F(\boldsymbol{x}_q^t - \boldsymbol{x}_r^t), \tag{3.56}$$

where $F \in [0, 2]$ is a parameter, often referred to as the differential weight. This requires that the minimum number of population size is $n \geq 4$. In principle, $F \in [0, 2]$, but in practice, a scheme with $F \in [0, 1]$ is more efficient and stable. In fact, almost all the studies in the literature use $F \in (0, 1)$.

Crossover is controlled by a crossover parameter $C_r \in [0, 1]$, controlling the rate or probability for crossover. The actual crossover can be carried out in two ways, binomial and exponential. The binomial scheme performs crossover on each of the D components or variables/parameters. By generating a uniformly distributed random number $r_i \in [0, 1]$ the jth component of $\boldsymbol{v}_i$ is manipulated as

$$\boldsymbol{u}_{j,i}^{t+1} = \begin{cases} \boldsymbol{v}_{j,i} & \text{if } r_i \leq C_r, \\ \boldsymbol{x}_{j,i}^t & \text{otherwise,} \end{cases} \qquad j = 1, 2, \ldots, D. \tag{3.57}$$

This way, each component can be decided randomly whether or not to exchange with the counterpart of the donor vector.

In the exponential scheme, a segment of the donor vector is selected, and this segment starts with random integer k and random length L, which can include many components. Mathematically, choosing $k \in [0, D-1]$ and $L \in [1, D]$ randomly, we have

$$\boldsymbol{u}_{j,i}^{t+1} = \begin{cases} \boldsymbol{v}_{j,i}^t & \text{for } j = k, \ldots, k-L+1 \in [1, D], \\ \boldsymbol{x}_{j,i}^t & \text{otherwise.} \end{cases} \tag{3.58}$$

The binomial scheme is simpler to implement.

Selection is essentially the same as that used in genetic algorithms. It is selecting the most fittest, that is, the minimum objective value for a minimization problem. Therefore we have

$$x_i^{t+1} = \begin{cases} u_i^{t+1} & \text{if } f(u_i^{t+1}) \leq f(x_i^t), \\ x_i^t & \text{otherwise.} \end{cases} \tag{3.59}$$

It is worth pointing out here that the use of $v_i^{t+1} \neq x_i^t$ may increase the evolutionary or exploratory efficiency. The overall search efficiency is controlled by two parameters, the differential weight F and the crossover probability C_r.

3.5.3 Particle swarm optimization

Many swarms in nature such as fish and birds can have higher-level behavior, but they all obey simple rules. For example, a swarm of birds such as starlings simply follow three basic rules: each bird flies according to the flight velocities of their neighbor birds (usually about seven adjacent birds) while keeping a certain separation distance; birds on the edge of the swarm tend to fly into the center of the swarm (so as to avoid being eaten by potential predators such as eagles); and, in addition, birds tend to fly to search for food or shelters, and thus a short memory is used. Based on such swarming characteristics, in 1995 particle swarm optimization (PSO) was developed by Kennedy and Eberhart [87], which uses equations to simulate the swarming characteristics of birds and fish.

For the ease of discussions, let us use x_i and v_i to denote the position (solution) and velocity, respectively, of a particle or agent i. In PSO, there are n particles as a population, and thus $i = 1, 2, \ldots, n$. There are two equations for updating positions and velocities of particles, and they can be written as follows:

$$v_i^{t+1} = v_i^t + \alpha \epsilon_1 [g^* - x_i^t] + \beta \epsilon_2 [x_i^* - x_i^t], \tag{3.60}$$

$$x_i^{t+1} = x_i^t + v_i^{t+1} \Delta t, \tag{3.61}$$

where ϵ_1 and ϵ_2 are two uniformly distributed random numbers in [0,1]. The learning parameters α and β are usually in the range of [0,2]. In Eq. (3.60), g^* is the best solution found so far by all the particles in the population, and each particle has an individual best solution x_i^* by itself during the entire past iteration history.

It is worth pointing out that $\Delta t = 1$ should be used because iterations in algorithms are discrete with a step counter $t \leftarrow t + 1$. Thus, there is no need to consider units and Δt in all algorithms discussed in this book.

3.5.4 Bat algorithm

Bat algorithm (BA), developed by Xin-She Yang [150,152] in 2010, uses some characteristics of frequency-tuning and echolocation of microbats. It also uses the variations of pulse emission rate r and loudness A to control exploration and exploitation. In the bat algorithm, main algorithmic equations for position $\boldsymbol{x}_i$ and velocity $\boldsymbol{v}_i$ for bat i are

$$f_i = f_{\min} + (f_{\max} - f_{\min})\beta, \tag{3.62}$$

$$\boldsymbol{v}_i^t = \boldsymbol{v}_i^{t-1} + (\boldsymbol{x}_i^{t-1} - \boldsymbol{x}_*) f_i, \tag{3.63}$$

$$\boldsymbol{x}_i^t = \boldsymbol{x}_i^{t-1} + \boldsymbol{v}_i^t \Delta t, \tag{3.64}$$

where $\beta \in [0, 1]$ is a random vector drawn from a uniform distribution so that the frequency can vary from $f_{\min}$ to $f_{\max}$. Here $\boldsymbol{x}_*$ is the current best solution found so far by all virtual bats. As pointed out earlier, $\Delta t = 1$ is used for iterative discrete algorithms.

From these equations we can see that both equations are linear in terms of $\boldsymbol{x}_i$ and $\boldsymbol{v}_i$. But the control of exploration and exploitation is carried out by the variations of loudness $A(t)$ from a high value to a lower value and the emission rate r from a lower value to a higher value, that is,

$$A_i^{t+1} = \alpha A_i^t, \quad r_i^{t+1} = r_i^0 (1 - e^{-\gamma t}), \tag{3.65}$$

where $0 < \alpha < 1$ and $\gamma > 0$ are two parameters. As a result, the actual algorithm can have a weak nonlinearity. Consequently, BA can have a faster convergence rate in comparison with PSO.

3.5.5 Firefly algorithm

Based on the flashing characteristics of tropical firefly species, in 2008 Xin-She Yang [148,149] developed the firefly algorithm (FA). FA uses a nonlinear system by combining the exponential decay of light absorption and inverse-square law of light variation with distance. In the FA, the main algorithmic equation for the position $\boldsymbol{x}_i$ (as a solution vector to a problem) is

$$\boldsymbol{x}_i^{t+1} = \boldsymbol{x}_i^t + \beta_0 e^{-\gamma r_{ij}^2} (\boldsymbol{x}_j^t - \boldsymbol{x}_i^t) + \alpha\, \boldsymbol{\epsilon}_i^t, \tag{3.66}$$

where α is a scaling factor controlling the step sizes of the random walks, whereas γ is a scale-dependent parameter controlling the visibility of the fireflies (and thus search modes). In addition, β_0 is the attractiveness constant when the distance between two fireflies is zero (i.e., $r_{ij} = 0$). This system is a nonlinear system, which may lead to rich characteristics in terms of algorithmic behavior.

Since the brightness of a firefly is associated with the objective landscape with its position as the indicator, the attractiveness of a firefly seen by others, depending on their relative positions and relative brightness. Thus the beauty is in the eye of the beholder. Consequently, a pair comparison is needed for comparing all fireflies.

3.5.6 Cuckoo search

In the natural world, among 141 cuckoo species, 59 species engage the so-called obligate brood parasitism. These cuckoo species do not build their own nests, and they lay eggs in the nests of host birds such as warblers. In fact, there is an arms race between cuckoo species and host species, forming an interesting cuckoo-host species coevolution system.

Based the above characteristics, Xin-She Yang and Suash Deb [154–156] developed in 2009 the cuckoo search (CS) algorithm. CS uses a combination of both local and global search capabilities, controlled by a discovery probability p_a. There are two algorithmic equations in CS, and one equation is

$$x_i^{t+1} = x_i^t + \alpha s \otimes H(p_a - \epsilon) \otimes (x_j^t - x_k^t), \tag{3.67}$$

where x_j^t and x_k^t are two different solutions selected randomly by random permutation, $H(u)$ is the Heaviside function, ϵ is a random number drawn from a uniform distribution, and s is the step size. This step is primarily local, though it can become global search if s is large enough. However, the main global search mechanism is realized by the other equation with Lévy flights:

$$x_i^{t+1} = x_i^t + \alpha L(s, \lambda), \tag{3.68}$$

where the Lévy flights are simulated (or drawn random numbers) by drawing random numbers from the Lévy distribution

$$L(s, \lambda) \sim \frac{\lambda \Gamma(\lambda) \sin(\pi \lambda / 2)}{\pi} \frac{1}{s^{1+\lambda}} \quad (s \gg 0). \tag{3.69}$$

Here $\alpha > 0$ is the step size scaling factor. It is worth pointing out that we use "$\sim$" here to highlight the fact that the steps are drawn from the distribution on the right-hand side as a sampling technique.

3.5.7 Flower pollination algorithm

Flower pollination algorithm (FPA) is a population-based algorithm, developed by Xin-She Yang and his collaborators, based on the inspiration from the pollination characteristics of flowering plants [153,157,159]. FPA intends to mimic some key characteristics of biotic and abiotic pollination as well as coevolutionary flower constancy between certain flower species and some pollinator species such as insects and animals.

In essence, there are two main equations for this algorithm, and the global search is carried out by

$$x_i^{t+1} = x_i^t + \gamma L(\lambda)(g_* - x_i^t), \tag{3.70}$$

where γ is a scaling parameter, $L(\lambda)$ is the random number vector drawn from a Lévy distribution governed by the exponent λ in the same form given in (3.69). Here g_* is the best solution found so far, which acts as a selection mechanism. The current solution x_i^t is modified by varying step sizes because Lévy flights can have a fraction of large step sizes in addition to many small steps.

The local search is carried out by

$$x_i^{t+1} = x_i^t + U(x_j^t - x_k^t), \tag{3.71}$$

which mimics local pollination and flower constancy. Here U is a uniformly distributed random number. Furthermore, x_j^t and x_k^t are solutions representing pollen from different flower patches.

As we mentioned earlier, the literature is expanding, and more nature-inspired algorithms are being developed by researchers, but we will not introduce more algorithms here. We refer the interested readers to more specialized literature such as Yang's book [159].

3.6 Notes on software

As we mentioned earlier, many software packages and programming languages have implemented some optimization capabilities, whereas commercial software packages tend to have well-tested toolboxes. It is not our intention to provide a comprehensive list of toolboxes and functionalities; we only intend to provide some flavor and diversity of a few software packages or programming languages.

- Matlab: The optimization toolboxes of Matlab include linear programming `linprog`, integer programming `intlinprog`, nonlinear programming such as `fminsearch` and `fmincon`, quadratic programming `quadprog`, and multiobjective optimization by genetic algorithm `gamultiobj`.
- Octave has many functionalities similar to Matlab, and it is an open-source package. Its optimization toolbox `optim` has implemented linear programming, quadratic programming, nonlinear programming, and linear least squares.
- R has a relatively general purpose optimization solver `optimr` with `optim()` using conjugate gradient, Nelder–Mead method, Broyden–Fletcher–Goldfarb–Shanno (BFGS) method, and simulated annealing. It also has a quadratic programming `solve.QP()` and least-squares solver `solve.qr()` as well as metaheuristic optimization such as the firefly algorithm.
- Python does have good optimization capabilities via `scipy.optimize()`, which includes the BFGS method, conjugate gradient, Newton's method, trust-region method, and least-square minimization.
- Mathematica is a commercial symbolic computation package. It has powerful functionalities for optimization, including nonlinear constrained global optimization `NMiminize` or `NMaximize`, linear programming and integer programming `LinearProgramming`, Knapsack solver `KnapsackSolve`, traveling salesman problem `FindShortesTour`, and others.
- Maple is mainly a symbolic and numerical computing tool with some functions for optimization such as `Minimize`, linear programming `LPSolve`, and nonlinear programming `NLPSolve`.
- Microsoft Excel Solver can do linear programming, integer programming, generalized reduced gradient, evolutionary algorithms (via a variant of the genetic

algorithm). On the other hand, the OpenSolver is free and has no limit on the number of variables. Its core algorithmic engine is COIN-OR linear and integer programming optimizers, and thus OpenSolver is very powerful and efficient.

Other powerful optimization tools include the computational infrastructure for Operations Research (COIN-OR), also known as common optimization interface for OR. Many software packages use it as a core optimization engine.

There are some Matlab demo codes for most of the nature-inspired algorithms discussed in this book. They are available from Matlab file exchanges,[2] including

- accelerated particle swarm optimization,[3]
- firefly algorithm,[4]
- cuckoo search,[5]
- flower pollination algorithm.[6]

It is worth pointing out that these codes are demo and incomplete codes. The reason is that such demo codes focus on the essential steps of the algorithms without any messy implementation of handling constraints. However, the performance of such concentrated demo codes may be reduced as the proper constraint-handling is an important part of practical applications. These codes should still work reasonably well for solving function optimization problems. This gives the readers an opportunity to understand the basic algorithms and potentially improve them.

[2] https://uk.mathworks.com/matlabcentral/profile/authors/2652824-xin-she-yang.

[3] https://uk.mathworks.com/matlabcentral/fileexchange/29725-accelerated-particle-swarm-optimization.

[4] https://uk.mathworks.com/matlabcentral/fileexchange/29693-firefly-algorithm.

[5] https://uk.mathworks.com/matlabcentral/fileexchange/29809-cuckoo-search-cs-algorithm.

[6] https://uk.mathworks.com/matlabcentral/fileexchange/45112-flower-pollination-algorithm.

Data fitting and regression

4

Contents

Regression can help to identify the trends in data and relationship between different quantities. Regression is one of the simplest forms of classification and supervised learning, and it is one of the most widely used data-processing techniques.

4.1 Sample mean and variance

If a sample consists of n independent observations $x_1, x_2, \ldots, x_n$ on a random variable x such as the noise level on a road or the price of a cup of coffee, two important and commonly used parameters are sample mean and sample variance, which can easily be estimated from the sample. The sample mean is calculated by

$$\bar{x} \equiv <x> = \frac{1}{n}(x_1 + x_2 + \cdots + x_n) = \frac{1}{n}\sum_{i=1}^{n} x_i, \tag{4.1}$$

which is in fact the arithmetic average of the values x_i.

The sample variance S^2 is defined by

$$S^2 = \frac{1}{n-1}\sum_{i=1}^{n}(x_i - \bar{x})^2. \tag{4.2}$$

Let us look at an example.

Introduction to Algorithms for Data Mining and Machine Learning. https://doi.org/10.1016/B978-0-12-817216-2.00011-9

Example 15

The measurements of a quantity such as the noise level on a road. The readings in dB are:

$$66,\ 73,\ 73,\ 74,\ 83,\ 70,\ 69,\ 77,\ 72,\ 75.$$

From the data we know that $n = 10$ and the mode is 73 as 73 appears twice (all the rest only appears once). The sample mean is

$$\begin{aligned}\bar{x} &= \frac{1}{10}(x_1 + x_2 + \cdots + x_{10}) \\ &= \frac{1}{10}(66 + 73 + 73 + 74 + 83 + 70 + 69 + 77 + 72 + 75) = \frac{732}{10} = 73.2.\end{aligned}$$

The corresponding sample variance is

$$\begin{aligned}S^2 &= \frac{1}{n-1}\sum_{i=1}^{n}(x_i - \bar{x})^2 \\ &= \frac{1}{10-1}\sum_{i=1}^{10}(x_i - 73.2)^2 \\ &= \frac{1}{9}[(66 - 73.2)^2 + (73 - 73.2)^2 + \cdots + (75 - 73.2)^2] \\ &= \frac{1}{9}[(-7.2)^2 + (-0.2)^2 + \cdots + (1.8)^2] = \frac{195.6}{9} \approx 21.73.\end{aligned}$$

Thus, the standard derivation is

$$S = \sqrt{S^2} \approx \sqrt{21.73} \approx 4.662.$$

Generally speaking, if u is a linear combination of n independent random variables $y_1, y_2, \ldots, y_n$ and each random variable y_i has an individual mean μ_i and a variance σ_i^2, we have the linear combination

$$u = \sum_{i=1}^{n} \alpha_i y_i = \alpha_1 y_1 + \alpha_2 y_2 + \cdots + \alpha_n y_n, \tag{4.3}$$

where the parameters α_i $(i = 1, 2, \ldots, n)$ are the weighting coefficients. The mean μ_u of the linear combination can be calculated by

$$\mu_u = E[u] = E[\sum_{i=1}^{n} \alpha_i y_i] = \sum_{i=1}^{n} \alpha_i E[y_i] = \sum \alpha_i \mu_i. \tag{4.4}$$

Then the variance σ_u^2 of the combination is

$$\sigma_u^2 = E[(u - \mu_u)^2] = E\Big[\sum_{i=1}^{n} \alpha_i (y_i - \mu_i)^2\Big], \tag{4.5}$$

which can be expanded as

$$\sigma_u^2 = \sum_{i=1}^{n} \alpha_i^2 E[(y_i - \mu_i)^2] + \sum_{i,j=1; i \neq j}^{n} \alpha_i \alpha_j E[(y_i - \mu_i)(y_j - \mu_j)], \tag{4.6}$$

where $E[(y_i - \mu_i)^2] = \sigma_i^2$. Since y_i and y_j are independent, we have

$$E[(y_i - \mu_i)(y_j - \mu_j)] = E[(y_i - \mu_i)]E[(y_j - \mu_j)] = 0. \tag{4.7}$$

Therefore we get

$$\sigma_u^2 = \sum_{i=1}^{n} \alpha_i^2 \sigma_i^2. \tag{4.8}$$

The sample mean defined in (4.1) can also be viewed as a linear combination of all the x_i assuming that each has the same mean $\mu_i = \mu$, variance $\sigma_i^2 = \sigma^2$, and weighting coefficient $\alpha_i = 1/n$. Hence the sample mean is an unbiased estimate of the sample due to the fact $\mu_{\bar{x}} = \sum_{i=1}^{n} \mu/n = \mu$. In this case, however, we have the variance

$$\sigma_{\bar{x}}^2 = \sum_{i=1}^{n} \frac{1}{n^2} \sigma^2 = \frac{\sigma^2}{n}, \tag{4.9}$$

which means that the variance becomes smaller as the size n of the sample increases by a factor of $1/n$.

For the sample variance S^2 defined earlier by

$$S^2 = \frac{1}{n-1} \sum_{i=1}^{n} (x_i - \bar{x})^2, \tag{4.10}$$

we can see that the factor is $1/(n-1)$, not $1/n$, because only $1/(n-1)$ will give the correct unbiased estimate of the variance. The other way to think about the factor $1/(n-1)$ is that we need at least one value to estimate the mean, and we need at least two values to estimate the variance. Thus, for n observations, only $n-1$ different values of variance can be obtained to estimate the total sample variance.

4.2 Regression analysis

Regression is a class of methods that are mostly based on the method of least squares and the maximum likelihood theory.

4.2.1 Maximum likelihood

For a sample of n values $y_1, y_2, \ldots, y_n$ of a random variable Y whose probability density function $p(y)$ depends on a set of k parameters $\beta_1, \ldots, \beta_k$, the joint probability

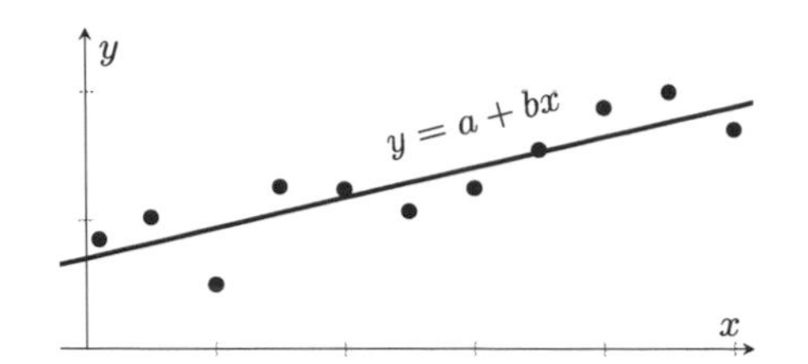

Figure 4.1 Best fit line for a simple linear model.

is the product of all the probabilities, that is,

$$\Phi(\beta_1, \dots, \beta_k) = \prod_{i=1}^{n} p(y_i, \beta_1, \dots, \beta_k)$$
$$= p(y_1, \beta_1, \dots, \beta_k) \cdot p(y_2, \beta_1, \dots, \beta_k) \cdots p(y_n, \beta_1, \dots, \beta_k), \tag{4.11}$$

where Π_i means the product of all its components. For example, $\Pi_{n=1}^{3} a_i = a_1 \times a_2 \times a_3$. The essence of the maximum likelihood is to maximize Φ by choosing the parameters β_j. As the sample can be considered as given values, the maximum likelihood requires the following stationarity conditions:

$$\frac{\partial \Phi}{\partial \beta_j} = 0 \quad (j = 1, 2, \dots, k), \tag{4.12}$$

whose solutions for β_j are the maximum likelihood estimates.

Regression is a particular case of the method of least-squares. Many other problems can be reformulated in this framework.

4.2.2 Liner regression

For experiments and observations, we usually plot one variable such as price or pressure y against another variable x such as time or spatial coordinates. We try to present the data in such a way that we can see some trend in the data.

For a set of n data points (x_i, y_i) $(i = 1, 2, \dots, n)$, the usual practice is to try to draw a straight line $y = a + bx$ so that it represents the major trend. Such a line is often called the regression line or the best fit line as shown in Fig. 4.1.

The method of linear least squares is to try to determine the two parameters, a (intercept) and b (slope), for the regression line from n data points, assuming that x_i are known more precisely and the values of y_i obey a normal distribution around the potentially best fit line with variance σ^2. So we have the joint probability with each being normally distributed:

$$P = \prod_{i=1}^{n} p(y_i) = A \exp\Big\{ -\frac{1}{2\sigma^2} \sum_{i=1}^{n} [y_i - f(x_i)]^2 \Big\}, \tag{4.13}$$

where A is a constant, and $f(x)$ is the function for the regression [$f(x)=a+bx$ for the linear regression].

It is worth pointing out that the exponent

$$\psi=\sum_{i=1}^{n}[y_i-f(x_i)]^2/(2\sigma^2) \tag{4.14}$$

is in fact a weighted sum of residuals or errors.

The maximization of P is equivalent to the minimization of ψ. To minimize ψ as a function of a and b via the model of $f(x)=a+bx$, its derivatives should be zero, that is,

$$\frac{\partial\psi}{\partial a}=-\frac{1}{\sigma^2}\sum_{i=1}^{n}[y_i-(a+bx_i)]=0 \tag{4.15}$$

and

$$\frac{\partial\psi}{\partial b}=-\frac{1}{\sigma^2}\sum_{i=1}^{n}x_i[y_i-(a+bx_i)]=0, \tag{4.16}$$

Since $\sigma^2\neq 0$, we can omit this factor, and these equations become

$$\sum_{i=1}^{n}[y_i-(a+bx_i)]=0,\quad \sum_{i=1}^{n}x_i[y_i-(a+bx_i)]=0. \tag{4.17}$$

By expanding these equations we have

$$na+b\sum_{i=1}^{n}x_i=\sum_{i=1}^{n}y_i \tag{4.18}$$

and

$$a\sum_{i=1}^{n}x_i+b\sum_{i=1}^{n}x_i^2=\sum_{i=1}^{n}x_iy_i, \tag{4.19}$$

which is a system of linear equations for a and b, and it is straightforward to obtain the solutions as

$$a=\frac{1}{n}[\sum_{i=1}^{n}y_i-b\sum_{i=1}^{n}x_i]=\bar{y}-b\bar{x}, \tag{4.20}$$

$$b=\frac{n\sum_{i=1}^{n}x_iy_i-(\sum_{i=1}^{n}x_i)(\sum_{i=1}^{n}y_i)}{n\sum_{i=1}^{n}x_i^2-(\sum_{i=1}^{n}x_i)^2}, \tag{4.21}$$

where

$$\bar{x} = \frac{1}{n}\sum_{i=1}^{n} x_i, \qquad \bar{y} = \frac{1}{n}\sum_{i=1}^{n} y_i. \tag{4.22}$$

If we use the following notations:

$$K_x = \sum_{i=1}^{n} x_i, \qquad K_y = \sum_{i=1}^{n} y_i, \tag{4.23}$$

and

$$K_{xx} = \sum_{i=1}^{n} x_i^2, \qquad K_{yy} = \sum_{i=1}^{n} y_i^2, \qquad K_{xy} = \sum_{i=1}^{n} x_i y_i, \tag{4.24}$$

then the equations for a and b give

$$a = \frac{K_{xx}K_y - K_x K_{xy}}{nK_{xx} - (K_x)^2} \tag{4.25}$$

and

$$b = \frac{nK_{xy} - K_x K_y}{nK_{xx} - (K_x)^2}. \tag{4.26}$$

The residual error is defined by

$$\epsilon_i = y_i - (a + bx_i), \tag{4.27}$$

whose sample mean is given by

$$\mu_\epsilon = \frac{1}{n}\sum_{i=1}^{n} \epsilon_i = \frac{1}{n} y_i - a - b\frac{1}{n}\sum_{i=1}^{n} x_i = \bar{y} - a - b\bar{x} = 0. \tag{4.28}$$

The sample variance S^2 is

$$S^2 = \frac{1}{(n-2)}\sum_{i=1}^{n}[y_i - (a + bx_i)]^2 = \frac{1}{(n-2)}\,\text{RSS}, \tag{4.29}$$

where the RSS stands for the residual sum of squares given by

$$\text{RSS} = \sum_{i=1}^{n}[y_i - f(x_i)]^2 = \sum_{i=1}^{n}[y_i - (a + bx_i)]^2. \tag{4.30}$$

Here the factor $1/(n-2)$ comes from the fact that two constraints are needed for the best fit, and therefore the residuals have $n-2$ degrees of freedom.

The correlation coefficient $r_{x,y}$ is a very useful parameter for finding any potential relationship between two sets of data x_i and y_i for two random variables x and y, respectively. If x has a mean $\bar{x}$ and a sample variance S_x^2 and if y has a mean $\bar{y}$ and a sample variance S_y^2, we have

$$\mathrm{var}(x) = S_x^2 = \frac{\sum_{i=1}^n (x_i - \bar{x})^2}{n-1}, \quad \mathrm{var}(y) = S_y^2 = \frac{\sum_{i=1}^n (y_i - \bar{y})^2}{n-1}. \tag{4.31}$$

The correlation coefficient is defined by

$$r_{x,y} = \frac{\mathrm{cov}(x,y)}{S_x S_y} = \frac{E[xy] - \bar{x}\bar{y}}{S_x S_y}, \tag{4.32}$$

where

$$\mathrm{cov}(x,y) = E[(x-\bar{x})(y-\bar{y})] = E[xy] - \bar{x}\bar{y} \tag{4.33}$$

is the covariance, which can be calculated explicitly by

$$\mathrm{cov}(x,y) = \frac{\sum_{i=1}^n (x_i - \bar{x})(y_i - \bar{y})}{n-1}. \tag{4.34}$$

It is worth pointing out that S_x and S_y must be sample variances; otherwise, the result is incorrect. In addition, we can also write

$$S_x = \mathrm{cov}(x,x), \quad S_y = \mathrm{cov}(y,y). \tag{4.35}$$

It is obvious that $\mathrm{cov}(x,y) = \mathrm{cov}(y,x)$. Thus, the covariance matrix

$$\boldsymbol{C}_s = \begin{pmatrix} \mathrm{cov}(x,x) & \mathrm{cov}(x,y) \\ \mathrm{cov}(y,x) & \mathrm{cov}(y,y) \end{pmatrix} = \begin{pmatrix} S_x & \mathrm{cov}(x,y) \\ \mathrm{cov}(x,y) & S_y \end{pmatrix} \tag{4.36}$$

is also symmetric.

If the two variables are independent or $\mathrm{cov}(x,y) = 0$, then there is no correlation between them ($r_{x,y} = 0$). If $r_{x,y}^2 = 1$, then there is a linear relationship between these two variables; $r_{x,y} = 1$ is an increasing linear relationship where the increase of one variable leads to the increase of the other. On the other hand, $r_{x,y} = -1$ is a decreasing relationship when one increases and the other decreases. In general, we have $-1 \le r_{x,y} \le 1$ or $|r_{x,y}| \le 1$.

For a set of n data points (x_i, y_i), the correlation coefficient can be calculated directly by

$$r_{x,y} = \frac{n \sum_{i=1}^n x_i y_i - \sum_{i=1}^n x_i \sum_{i=1}^n y_i}{\sqrt{\left[n \sum x_i^2 - (\sum_{i=1}^n x_i)^2\right]\left[n \sum_{i=1}^n y_i^2 - (\sum_{i=1}^n y_i)^2\right]}}$$

or

$$r_{x,y} = \frac{n K_{xy} - K_x K_y}{\sqrt{(n K_{xx} - K_x^2)(n K_{yy} - K_y^2)}}, \tag{4.37}$$

where $K_{yy} = \sum_{i=1}^n y_i^2$.

Table 4.1 Two quantities with measured data.

Input (*H*)	Output (*Y*)	Input (*H*)	Output (*Y*)
90	270	300	910
110	330	350	1080
140	410	400	1270
170	520	450	1450
190	560	490	1590
225	670	550	1810
250	750	650	2180

Now let us look at an example that are measurements of two quantities (H and Y). The data for a set of random samples are given in Table 4.1.

Is there any relationship between these two quantities?

Now let us try to do a linear regression in the following form:

$$Y = a + bH.$$

Example 16

From the data in Table 4.1 with $n = 14$ we can calculate

$$K_H = \sum_{i=1}^{14} H_i = 90 + 110 + \cdots + 650 = 4365,$$

$$K_Y = \sum_{i=1}^{14} Y_i = 270 + 330 + \cdots + 2180 = 13800,$$

$$K_{HY} = \sum_{i=1}^{14} H_i Y_i = 90 * 270 + \cdots + 650 * 2180 = 5654150,$$

$$K_{HH} = \sum_{i=1}^{14} H_i^2 = 90^2 + \cdots + 650^2 = 1758025,$$

and

$$K_{YY} = \sum_{i=1}^{14} Y_i^2 = 270^2 + \cdots + 2180^2 = 18211800.$$

Thus we get

$$a = \frac{K_{HH} K_Y - K_H K_{HY}}{n K_{HH} - K_H^2}$$
$$= \frac{1758025 \times 13800 - 4365 \times 5654150}{14 \times 1758025 - 4365^2} \approx -75.48$$

and

$$b = \frac{nK_{HY} - K_H K_Y}{nK_{HH} - K_H^2}$$
$$= \frac{14 \times 5654150 - 4365 \times 13800}{14 \times 1758025 - 4365^2} \approx 3.404.$$

So the regression line becomes

$$Y = -75.48 + 3.404H.$$

Therefore, their correlation coefficient r is given by

$$r = \frac{nK_{HY} - K_H K_Y}{\sqrt{(nK_{HH} - K_H^2)(nK_{YY} - K_Y^2)}}$$
$$= \frac{14 \times 5654150 - 4365 \times 13800}{\sqrt{(14 \times 1758025 - 4365^2)(14 \times 18211800 - 13800^2)}} \approx 0.99903.$$

This is indeed a relatively strong correlation.

These formulations are based on the fact that the curve-fitting function $y = f(x) = a + bx$ is linear in terms of the independent variable x and the parameters (a and b). Here the key linearity is about parameters but not about the basis function x. Thus, the previous technique can still be applicable to both functions $f(x) = a + bx + cx^2$ and $g(x) = a + b\sin x$ with some minor adjustments to be discussed later in this chapter. However, if we have a function of the form

$$y = \ln(a + bx),$$

then the technique cannot be applied directly, and some linearization approximations should be used.

4.2.3 Linearization

Sometimes, some obviously nonlinear functions can be transformed into linear forms so as to carry out linear regression, instead of more complicated nonlinear regression. However, there is no general formula for such linearization, and thus it is often necessary to deal with each case individually. This can be illustrated by some examples.

Example 17

For example, the nonlinear function

$$f(x) = \alpha e^{-\beta x} \tag{4.38}$$

can be transformed into a linear form by taking logarithms of both sides. We have

$$\ln f(x) = \ln(\alpha) - \beta x, \tag{4.39}$$

which is equivalent to $y = a + bx$ if we let $y = \ln f(x)$, $a = \ln(\alpha)$, and $b = -\beta$.

In addition, the function

$$f(x) = \alpha e^{-\beta x+\gamma} = Ae^{-\beta x},$$

where $A = \alpha e^{\gamma}$ is essentially the same as the previous function.

Similarly, function

$$f(x) = \alpha x^{\beta} \tag{4.40}$$

can also be transformed into

$$\ln[f(x)] = \ln(\alpha) + \beta \ln(x), \tag{4.41}$$

which is a linear regression $y = a + b\zeta$ between $y = \ln[f(x)]$ and $\zeta = \ln(x)$, where $a = \ln(\alpha)$ and $b = \beta$.

Furthermore, the function

$$f(x) = \alpha\beta^{x} \tag{4.42}$$

can also be converted into the standard linear form

$$\ln f(x) = \ln\alpha + x\ln\beta \tag{4.43}$$

by letting $y = \ln[f(x)]$, $a = \ln\alpha$, and $b = \ln\beta$.

It is worth pointing out that the data points involving zeros should be taken out due to the potential singularity of the logarithm. Fortunately, these points rarely occur in the regression for the functions of the forms mentioned.

Example 18

If a set of data can fit to the nonlinear function

$$y = ax\exp(-x/b)$$

in the range of $(0, \infty)$, it is then possible to convert it to a linear regression.

As $x = 0$ is just a single point, we can leave this out. For $x \neq 0$, we can divide both sides by x:

$$\frac{y}{x} = a\exp(-x/b).$$

Taking the logarithm of both sides, we have

$$\ln\frac{y}{x} = \ln a - \frac{1}{b}x,$$

which is a linear regression of y/x versus x.

In general, linearization is possible for only a small class of nonlinear functions. For nonlinear functions, we have to use either approximation or full nonlinear least squares to be introduced later in this chapter.

4.2.4 Generalized linear regression

The most widely used linear regression is the so-called generalized least square as a linear combination of basis functions. Fitting to a polynomial of degree p,

$$y(x) = \alpha_0 + \alpha_1 x + \alpha_2 x^2 + \cdots + \alpha_p x^p, \tag{4.44}$$

is probably the most widely used. This is equivalent to the regression to the linear combination of the basis functions $1, x, x, \ldots,$ and x^p. However, there is no particular reason why we have to use these basis functions. In fact, the basis functions can be any arbitrary known functions such as $\sin x$, $\cos x$ and even $\exp(x)$, and the main requirement is that the model can be explicitly expressed as a linear combination of basis functions. In this sense, the generalized least square can be written as

$$y(x) = \sum_{j=0}^{p} \alpha_j f_j(x), \tag{4.45}$$

where the basis functions f_j are known functions of x without any unknown or undetermined parameters.

Now the sum of least squares is defined as

$$\psi = \sum_{i=1}^{n} \frac{[y_i - \sum_{j=0}^{p} \alpha_j f_j(x_i)]^2}{\sigma_i^2}, \tag{4.46}$$

where σ_i $(i = 1, 2, \ldots, n)$ are the standard deviations of the ith data point at (x_i, y_i). There are n data points in total. To determine the coefficients uniquely, it is required that

$$n \geq p + 1. \tag{4.47}$$

In the case of unknown standard deviations σ_i, a common practice is to set all the values σ_i as the same constant $\sigma_i = \sigma$, which can be moved to the outside of the summation and thus be omitted as it will not affect the results.

Let $\boldsymbol{D} = [D_{ij}]$ be the design matrix given by

$$D_{ij} = \frac{f_j(x_i)}{\sigma_i}. \tag{4.48}$$

The minimum of ψ is determined by

$$\frac{\partial \psi}{\partial \alpha_j} = 0 \qquad (j = 0, 1, \ldots, p), \tag{4.49}$$

that is,

$$\sum_{i=1}^{n} \frac{f_k(x_i)}{\sigma_i^2} \Big[y_i - \sum_{j=0}^{p} \alpha_j f_j(x_i) \Big] = 0, \qquad k = 0, \ldots, p. \tag{4.50}$$

Rearranging the terms and interchanging the order of summations, we have

$$\sum_{j=0}^{p}\sum_{i=1}^{n}\frac{\alpha_j f_j(x_i) f_k(x_i)}{\sigma_i^2}=\sum_{i=1}^{n}\frac{y_i f_k(x_i)}{\sigma_i^2}, \tag{4.51}$$

which can be written compactly as the matrix equation

$$\sum_{j=0}^{p} A_{kj}\alpha_j = b_k, \tag{4.52}$$

or

$$\boldsymbol{A}\boldsymbol{\alpha} = \boldsymbol{b}, \tag{4.53}$$

where

$$\boldsymbol{A} = \boldsymbol{D}^T \boldsymbol{D}$$

is a $(p+1) \times (p+1)$ matrix, that is,

$$A_{kj} = \sum_{i=1}^{n}\frac{f_k(x_i) f_j(x_i)}{\sigma_i^2}. \tag{4.54}$$

Here $\boldsymbol{b} = [b_k]$ is the column vector given by

$$b_k = \sum_{i=1}^{n}\frac{y_i f_k(x_i)}{\sigma_i^2}, \tag{4.55}$$

where $k = 0, \dots, p$. Eq. (4.52) is a linear system of the so-called normal equations, which can be solved using the standard methods for solving linear systems. The solution of the coefficients is $\boldsymbol{\alpha} = \boldsymbol{A}^{-1}\boldsymbol{b}$ or

$$\alpha_k = \sum_{j=0}^{p}[A]_{kj}^{-1} b_j \qquad (k = 0, \dots, p), \tag{4.56}$$

where $\boldsymbol{A}^{-1} = [A]_{ij}^{-1}$.

A particular case of the generalized linear least squares is the so-called polynomial least squares when the basis functions are simple power functions $f_i(x) = x^i$ $(i = 0, 1, \dots, p)$, that is,

$$f_i(x) = 1,\ x,\ x^2, \dots,\ x^p. \tag{4.57}$$

For simplicity, we assume that $\sigma_i = \sigma = 1$. The matrix equation (4.52) simply becomes

$$\begin{pmatrix} \sum_{i=1}^n 1 & \sum_{i=1}^n x_i & \dots & \sum_{i=1}^n x_i^p \\ \sum_{i=1}^n x_i & \sum_{i=1}^n x_i^2 & \dots & \sum_{i=1}^n x_i^{p+1} \\ \vdots & & \ddots & \\ \sum_{i=1}^n x_i^p & \sum_{i=1}^n x_i^{p+1} & \dots & \sum_{i=1}^n x_i^{2p} \end{pmatrix} \begin{pmatrix} \alpha_0 \\ \alpha_1 \\ \vdots \\ \alpha_p \end{pmatrix} = \begin{pmatrix} \sum_{i=1}^n y_i \\ \sum_{i=1}^n x_i y_i \\ \vdots \\ \sum_{i=1}^n x_i^p y_i \end{pmatrix}.$$

In the simplest case where $p = 1$, it becomes the standard linear regression

$$y = \alpha_0 + \alpha_1 x = a + bx.$$

Now we have

$$\begin{pmatrix} n & \sum_{i=1}^n x_i \\ \sum_{i=1}^n x_i & \sum_{i=1}^n x_i^2 \end{pmatrix} \begin{pmatrix} \alpha_0 \\ \alpha_1 \end{pmatrix} = \begin{pmatrix} \sum_{i=1}^n y_i \\ \sum_{i=1}^n x_i y_i \end{pmatrix}. \tag{4.58}$$

Its solution is

$$\begin{pmatrix} \alpha_0 \\ \alpha_1 \end{pmatrix} = \frac{1}{\Delta} \begin{pmatrix} \sum_{i=1}^n x_i^2 & -\sum_{i-1}^n x_i \\ -\sum_{i=1}^n x_i & n \end{pmatrix} \begin{pmatrix} \sum_{i=1}^n y_i \\ \sum_{i=1}^n x_i y_i \end{pmatrix}$$
$$= \frac{1}{\Delta} \begin{pmatrix} (\sum_{i=1}^n x_i^2)(\sum_{i=1}^n y_i) - (\sum_{i=1}^n x_i)(\sum_{i=1}^n x_i y_i) \\ n \sum_{i=1}^n x_i y_i - (\sum_{i=1}^n x_i)(\sum_{i=1}^n y_i) \end{pmatrix}, \tag{4.59}$$

where

$$\Delta = n \sum_{i=1}^n x_i^2 - (\sum_{i=1}^n x_i)^2. \tag{4.60}$$

These are exactly the same coefficients as those in Eq. (4.26).

Example 19

We now use a quadratic function to best fit the following data (as shown in Fig. 4.2):

$$x: -0.98, \ 1.00, \ 2.02, \ 3.03, \ 4.00$$
$$y: \ 2.44, \ -1.51, \ -0.47, \ 2.54, \ 7.52$$

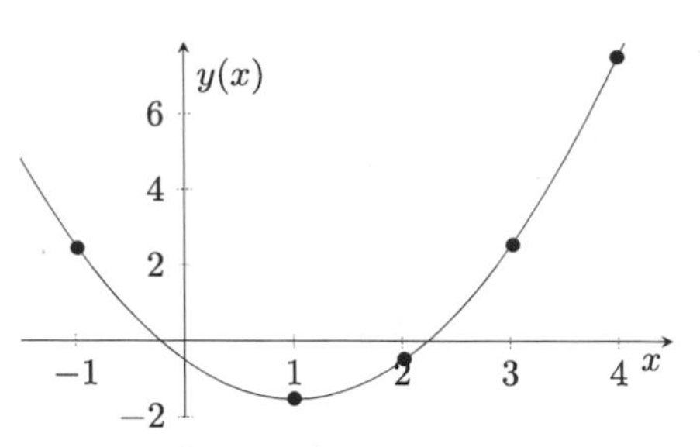

Figure 4.2 Best fit curve for $y(x) = x^2 - 2x - \frac{1}{2}$ with 2.5% noise.

For the formula $y = \alpha_0 + \alpha_1 x + \alpha_2 x^2$, we have

$$\begin{pmatrix} n & \sum_{i=1}^n x_i & \sum_{i=1}^n x_i^2 \\ \sum_{i=1}^n x_i & \sum_{i=1}^n x_i^2 & \sum_{i=1}^n x_i^3 \\ \sum_{i=1}^n x_i^2 & \sum_{i=1}^n x_i^3 & \sum_{i=1}^n x_i^4 \end{pmatrix} \begin{pmatrix} \alpha_0 \\ \alpha_1 \\ \alpha_2 \end{pmatrix} \begin{pmatrix} \sum_{i=1}^n y_i \\ \sum_{i=1}^n x_i y_i \\ \sum_{i=1}^n x_i^2 y_i \end{pmatrix}.$$

Using the data set, we have $n = 5$, $\sum_{i=1}^n x_i = 9.07$, and $\sum_{i=1}^n y_i = 10.52$. Other quantities can be calculated in a similar way. Therefore we have

$$\begin{pmatrix} 5.0000 & 9.0700 & 31.2217 \\ 9.0700 & 31.2217 & 100.119 \\ 31.2217 & 100.119 & 358.861 \end{pmatrix} \begin{pmatrix} \alpha_0 \\ \alpha_0 \\ \alpha_2 \end{pmatrix} = \begin{pmatrix} 10.52 \\ 32.9256 \\ 142.5551 \end{pmatrix}.$$

By direct inversion we have

$$\begin{pmatrix} \alpha_0 \\ \alpha_1 \\ \alpha_2 \end{pmatrix} = \begin{pmatrix} -0.5055 \\ -2.0262 \\ 1.0065 \end{pmatrix}.$$

Finally, the best fit equation is

$$y(x) = -0.5055 - 2.0262x + 1.0065x^2,$$

which is quite close to the formula $y = x^2 - 2x - 1/2$ used to generate the original data with a random component of about 2.5%. The total residual sum of squares (RSS) is RSS $= 0.0045$. The fit seems to be highly accurate.

4.2.5 Goodness of fit

In the above example, if we choose $p = 2$ (a quadratic polynomial), then the curve-fitting seems to work very well. But how do we know which order of polynomials to use in the first place? In fact, the degree p is a hyperparameter for this curve-fitting

Table 4.2 Goodness of fit in terms of RSS.

Order	$p = 1$	$p = 2$	$p = 3$	$p = 4$
RSS	36.3485	0.0045	0.0080	6.9×10^{-30}

problem, and we have to use some additional information to find the right value for this parameter.

Suppose, we start with $p = 1$ (a straight line) and carry out the regression in the similar way as we before. We should get a best-fit line

$$f_1(x) = 0.9373x + 0.4038, \tag{4.61}$$

with the RSS $= 36.3485$.

If we use $p = 3$, then we have

$$f_3(x) = 0.0080x^3 + 0.9694x^2 - 2.0131x - 0.4580, \tag{4.62}$$

with the RSS $= 0.002$, which is the smallest RSS for $p = 1, 2, 3$.

If we used RSS as the goodness of fit, then it seems $p = 3$ gives a better fit than $p = 2$, even though the coefficient of the highest order x^3 is 0.0080. Now if we proceed this way, what happens if we use $p = 4$?

If we use $p = 4$, then we have

$$f_4(x) = 0.0101x^4 - 0.0610x^3 + 1.0778x^2 - 1.9571x - 0.5798, \tag{4.63}$$

with even a smaller RSS $= 6.9 \times 10^{-30}$. We summarize our results in Table 4.2.

However, we cannot continue this way because we do not have enough data to produce well-posed coefficients if p is higher than n. In general, as p increases, the RSS of the data points can usually decrease, but the oscillations between data points can increase dramatically.

In addition, higher-order models may introduce unrealistic model parameters not supported by the data. This is the well-known overfitting phenomenon, which should be avoided. We will discuss some approaches such as information criteria and regularization to deal with overfitting in later sections.

4.3 Nonlinear least squares

As functions in most mathematical models are nonlinear, we need the nonlinear least squares in general. For given n data points (x_i, y_i) $(i = 1, 2, \ldots, n)$, we can fit a model $f(x_i, \boldsymbol{a})$ where $\boldsymbol{a} = (a_0, a_1, \ldots, a_m)^T$ is a vector of $m + 1$ parameters. In the simplest linear case, we have $f(x_i, \boldsymbol{a}) = a_0 + a_1 x$. For the one-variable logistic regression to be discussed later, we have

$$f(x_i, \boldsymbol{a}) = 1/(1 + e^{a_0 + a_1 x}). \tag{4.64}$$

In general, we have the nonlinear least squares to minimize the L_2-norm of the residuals $R_i = y_i - f(x_i, \boldsymbol{a})$, that is, to minimize the fitting error:

$$\text{minimize } E(\boldsymbol{a}) = \sum_{i=1}^{n} R_i^2(\boldsymbol{a}) = \sum_{i=1}^{n} [y_i - f(x_i, \boldsymbol{a})]^2 = ||R_i(\boldsymbol{a})||_2^2. \tag{4.65}$$

If we treat this as an optimization problem so as to find the best $\boldsymbol{a}$, then we can use any optimization techniques such as Newton's method to solve this optimization problem. However, we can use the properties of this problem to get the solution more efficiently [136].

4.3.1 Gauss–Newton algorithm

Let $\boldsymbol{J}$ denote the Jacobian matrix of the form

$$\boldsymbol{J} = [J_{ij}] = \frac{\partial R_i}{\partial a_j}, \quad i = 1, 2, \ldots, n, \quad j = 0, 1, 2, \ldots, m, \tag{4.66}$$

which is an $n \times (m+1)$ matrix. Then the gradient of the objective (error) function is the differentiation of E:

$$\frac{\partial E}{\partial a_j} = 2 \sum_{i=1}^{n} \frac{\partial R_i}{\partial a_j} R_i, \quad j = 0, 1, 2, \ldots, m, \tag{4.67}$$

which can be written in the vector form as

$$\nabla E(\boldsymbol{a}) = 2 \boldsymbol{J}^T \boldsymbol{R}, \tag{4.68}$$

where the residual vector $\boldsymbol{R}$ is given by

$$\boldsymbol{R}(\boldsymbol{a}) = \Big(\; R_1(\boldsymbol{a}), R_2(\boldsymbol{a}), \ldots, R_n(\boldsymbol{a}) \; \Big)^T. \tag{4.69}$$

Here ∇E is a vector with $m+1$ components. Similarly, the Hessian matrix $\boldsymbol{H}$ of the objective E can be written as

$$\boldsymbol{H} = \nabla^2 E = 2 \sum_{i=1}^{n} [\nabla R_i \nabla R_i^T + R_i \nabla^2 R_i] = 2 \boldsymbol{J}^T \boldsymbol{J} + 2 \sum_{i=1}^{n} R_i \nabla^2 R_i, \tag{4.70}$$

which gives an $(m+1) \times (m+1)$ matrix. As it is expected that R_i will get smaller as the goodness of fit increases, we can essentially approximate the Hessian matrix by ignoring all the higher-order terms, so we have

$$\boldsymbol{H} \approx 2 \boldsymbol{J}^T \boldsymbol{J}. \tag{4.71}$$

Now we can solve the nonlinear least squares (4.65) by Newton's method, and we have

$$\begin{aligned}\boldsymbol{a}_{t+1} &= \boldsymbol{a}_t - \frac{\nabla E}{\nabla^2 E} \\ &= \boldsymbol{a}_t - \frac{2\boldsymbol{J}^T\boldsymbol{R}}{2\boldsymbol{J}^T\boldsymbol{J}} = \boldsymbol{a}_t - (\boldsymbol{J}^T\boldsymbol{J})^{-1}\boldsymbol{J}^T\boldsymbol{R}(\boldsymbol{a}_t).\end{aligned} \tag{4.72}$$

The initial vector $\boldsymbol{a}_0$ should be a good educated guess though $\boldsymbol{a}_0 = (1, 1, \ldots, 1)^T$ may work well in most cases.

Here we have used the approximation of $\boldsymbol{H}$ by $2\boldsymbol{J}^T\boldsymbol{J}$. In this case, this iterative method is often referred to the Gauss–Newton method, which usually has a good convergence rate. However, it is required that $\boldsymbol{J}$ should have a full rank so that the inverse $(\boldsymbol{J}^T\boldsymbol{J})^{-1}$ exists.

It is worth pointing out that in the context of line search $\boldsymbol{a}_{t+1} = \boldsymbol{a}_t + \alpha_t\boldsymbol{s}_t$, where $\boldsymbol{s}_t$ is the step size and $0 < \alpha_t \le 1$ is the scaling parameter or learning rate, this Gauss–Newton method is equivalent to a line search with a step size

$$\boldsymbol{J}^T\boldsymbol{J}\boldsymbol{s}_t = -\boldsymbol{J}^T\boldsymbol{R}(\boldsymbol{a}_t). \tag{4.73}$$

Though the iteration does not necessarily lead to the reduction of E in every iteration, it is better to choose α_t such that

$$E(\boldsymbol{a}_t + \alpha_t\boldsymbol{s}_t) < E(\boldsymbol{a}_t), \tag{4.74}$$

which leads to the reduction of least square errors.

Example 20

As an example, let us use the following data:

$$x : 0.10,\ 0.50,\ 1.0,\ 1.5,\ 2.0,\ 2.5$$
$$y : 0.10,\ 0.28,\ 0.40,\ 0.40,\ 0.37,\ 0.32$$

We fit the data to a model

$$y = \frac{x}{a + bx^2}, \tag{4.75}$$

where a and b are the coefficients to be determined by the data. The objective is to minimize the sum of the residual squares

$$S = \sum_{i=1}^{6} R_i^2 = \sum_{i=1}^{6} \Big[1 - \frac{x}{a + bx^2}\Big]^2, \tag{4.76}$$

where

$$R_i = y_i - \frac{x_i}{a + bx_i^2}.$$

We have

$$\frac{\partial R_i}{\partial a} = \frac{x_i}{(a + bx_i^2)^2}, \quad \frac{\partial R_i}{\partial b} = \frac{x_i^3}{(a + bx_i^2)^2}.$$

If we use the initial guess $a = 1$ and $b = 1$, then the initial residuals are

$$\boldsymbol{R} = \begin{pmatrix} 0.0010 & -0.1200 & -0.1000 & -0.0615 & -0.0300 & -0.0248 \end{pmatrix}^T.$$

The initial Jacobian matrix is

$$\boldsymbol{J} = \begin{pmatrix} 0.0980 & 0.0010 \\ 0.3200 & 0.0800 \\ 0.2500 & 0.2500 \\ 0.1420 & 0.3195 \\ 0.0800 & 0.3200 \\ 0.0476 & 0.2973 \end{pmatrix}.$$

Thus the first iteration using the Gauss–Newton algorithm gives

$$\begin{pmatrix} a \\ b \end{pmatrix}_1 = \begin{pmatrix} 1 \\ 1 \end{pmatrix} - (\boldsymbol{J}^T \boldsymbol{J})^{-1} \boldsymbol{J} \boldsymbol{R} = \begin{pmatrix} 1.3449 \\ 1.0317 \end{pmatrix}.$$

Then, updating the new Jacobian and residuals, we have

$$a = 1.4742 \quad \text{and} \quad b = 1.0059$$

after the second iteration. Similarly, we have

$$a = 1.4852, \ b = 1.0022 \ \text{(third iteration)},$$

and

$$a = 1.4854, \ b = 1.0021 \ \text{(fourth iteration)}.$$

In fact, this converges quickly, and the parameters almost remain the same values even after 10 iterations. The data points and the best fit model are shown in Fig. 4.3.

In general, the Guass–Newton method can work very well for a wide range of nonlinear curve fitting problems, even for large-scale problems. However, when the elements of the Jacobian are small (close to zeros), the matrix $\boldsymbol{J}^T \boldsymbol{J}$ may become singular, and thus the pseudo-inverse may become ill-posed. In addition, when approaching the optimality, the gradient becomes close to zero, and the convergence becomes very slow. A possible remedy is to use the Levenberg–Marquardt algorithm.

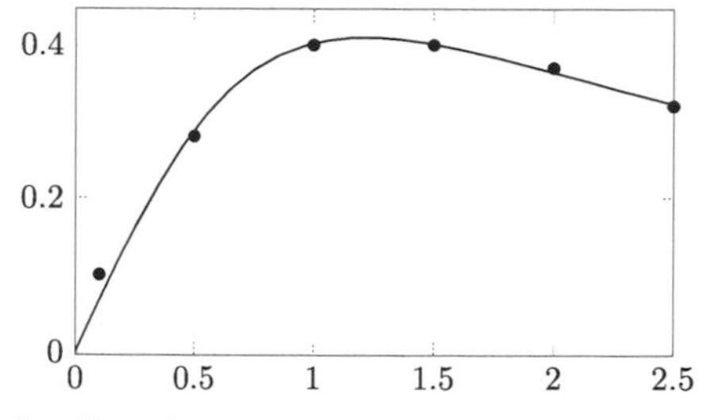

Figure 4.3 An example of nonlinear least squares.

4.3.2 Levenberg–Marquardt algorithm

In essence, the Levenberg–Marquardt algorithm is more robust by using a damping term in the approximation of the Hessian, that is,

$$\boldsymbol{H} \approx 2[\boldsymbol{J}^T \boldsymbol{J} + \mu \boldsymbol{I}], \tag{4.77}$$

where $\mu > 0$ is the damping coefficient, also called the Marquardt parameter, and $\boldsymbol{I}$ is the identity matrix of the same size as $\boldsymbol{H}$. Thus, the iteration formula becomes

$$\boldsymbol{a}_{t+1} = \boldsymbol{a}_t - \frac{\boldsymbol{J}^T \boldsymbol{R}(\boldsymbol{a}_t)}{\boldsymbol{J}^T \boldsymbol{J} + \mu \boldsymbol{I}} = \boldsymbol{a}_t - (\boldsymbol{J}^T \boldsymbol{J} + \mu \boldsymbol{I})^{-1} \boldsymbol{J}^T \boldsymbol{R}(\boldsymbol{a}_t), \tag{4.78}$$

which is equivalent to the step size $\boldsymbol{s}_t$ given by

$$(\boldsymbol{J}^T \boldsymbol{J} + \mu \boldsymbol{I}) \boldsymbol{s}_t = -\boldsymbol{J}^T \boldsymbol{R}(\boldsymbol{a}_t). \tag{4.79}$$

Mathematically speaking, a large μ effectively reduces the step size (in comparison with those in the Gauss–Newton algorithm) and damps the moves so that the descent is in the right direction with right amount. If the reduction in E is sufficient, then we can either keep this value of μ or reduce it. However, if the reduction E is not sufficient, then we can increase μ. Thus μ should vary as iteration continues, and there are various schemes for varying this hyperparameter. It is obvious that this method reduces to the standard Gauss–Newton algorithm if $\mu = 0$. Any nonzero μ essentially ensures that the matrix $\boldsymbol{J}^T \boldsymbol{J}$ is full rank, and thus the iterations can be more robust.

An alternative view of the Levenberg–Marquardt algorithm is the approximation of $\boldsymbol{H}$ in a trust region, which is equivalent to the minimization problem

$$\text{minimize} \quad ||\boldsymbol{J}^T \boldsymbol{s}_t + \boldsymbol{R}||_2^2, \quad |\boldsymbol{s}_t| \le \Delta_t, \tag{4.80}$$

where Δ_k is the radius of the trust region. For details, we refer the readers to the more advanced literature [32].

4.3.3 Weighted least squares

In many cases, the measurement errors in data may be different, or the assumption of equal variances in data may not be true. In this case, we need to weight the residuals differently, which leads to the so-called weighted least squares

$$\text{minimize} \sum_{i=1}^{n} w_i R_i^2 = \sum_{i=1}^{n} \frac{R_i^2}{\sigma_i^2} = ||\frac{R_i}{\sigma_i}||_2^2, \tag{4.81}$$

where

$$R_i = y_i - f(x_i, \boldsymbol{a}). \tag{4.82}$$

Here $w_i = 1/\sigma_i^2$ and σ_i^2 is the variance associated with data point (x_i, y_i).

By defining a weight matrix $\boldsymbol{W}$ as

$$\boldsymbol{W} = \text{diag}(w_i) = \begin{pmatrix} w_1 & 0 & \dots & 0 \\ 0 & w_2 & \dots & 0 \\ \vdots & \vdots & \ddots & \vdots \\ 0 & 0 & \dots & w_n \end{pmatrix}, \tag{4.83}$$

and following the similar derivations as before Eq. (4.72) becomes

$$\boldsymbol{a}_{t+1} = \boldsymbol{a}_t - (\boldsymbol{J}^T \boldsymbol{W} \boldsymbol{J})^{-1} (\boldsymbol{J}^T \boldsymbol{W} \boldsymbol{R}), \tag{4.84}$$

which is equivalent to approximating the Hessian by $\boldsymbol{H} = 2\boldsymbol{J}^T \boldsymbol{W} \boldsymbol{J}$ and the gradient by $\nabla E = 2\boldsymbol{J}^T \boldsymbol{W} \boldsymbol{R}$.

There are other regression methods. For example, for classification purposes, the logistic regression of the form

$$y(\boldsymbol{x}) = \frac{1}{1 + e^{-(a+bx)}} \tag{4.85}$$

is often used because its outputs can be interpreted as binary (0 or 1) classification. Another powerful regression method is the principal component analysis, which is a multivariate regression method. We will introduce these methods later in the next chapter. Now let us first discuss overfitting and information criteria.

4.4 Overfitting and information criteria

As we can see in the previous sections, curvefitting and regression are optimization in the least-squares sense because the objective is minimizing the fitting errors from the target values. In principle, the errors at known data points can become sufficiently small if higher-order polynomials are used; however, oscillations between data points become more severe. This leads to the so-called overfitting, which subsequently gives overcomplicated models. Ideally, the fitting should be guided by Occam's razor, which states that if there are many competing models to explain or fit the same data, then that with the fewest assumptions or parameters should be selected. The issue of overfitting is relevant to many applications such as curve-fitting, regression in general, and training of neural networks.

It is worth pointing out that the degree of polynomials to be used for curve fitting is a hyperparameter, which needs extra information or rule to determine. Though sophisticated methods in model selection may help, some simple criteria such as the Bayesian information criterion (BIC) or Akaike information criterion (AIC) can be used to select such hyperparameters to avoid overfitting [2].

For a statistical model, such as regression with k parameters, the Akaike information criterion (AIC) is defined by

$$\text{AIC} = 2k - 2\ln L, \tag{4.86}$$

where L is the maximum value of the likelihood function. For n data points with errors being independent identical normal distributions, we have

$$\text{AIC} = 2k + n\ln\Big(\frac{\text{RSS}}{n}\Big), \tag{4.87}$$

where RSS is the residual sum of squares, that is,

$$\text{RSS} = \sum_{i=1}^{n}[y_i - \hat{y}_i(x_i)]^2, \tag{4.88}$$

where y_i $(i = 1, 2, \ldots, n)$ are the true values, whereas $\hat{y}_i(x_i)$ $(i = 1, 2, \ldots, n)$ are the values predicted by the model. In principle, the minimization of AIC gives the best k.

However, this AIC may become inaccurate when the sample size is small; especially, when $n/k < 40$, we have to use a corrected AIC, called AICc given by

$$\begin{aligned} \text{AIC}_c &= \text{AIC} + \frac{2k(k+1)}{n-k-1} \\ &= 2k + n\ln\Big(\frac{\text{RSS}}{n}\Big) + \frac{2k(k+1)}{n-k-1}. \end{aligned} \tag{4.89}$$

In essence, the AIC is equivalent to the principle of maximum entropy.

Another information criterion is the Bayesian information criterion (BIC), which can be written as

$$\text{BIC} = k\ln n - 2\ln L. \tag{4.90}$$

With the same assumptions of errors obeying Gaussian distributions, we have

$$\text{BIC} = k\ln n + n\ln\Big(\frac{\text{RSS}}{n}\Big). \tag{4.91}$$

Both AIC and BIC are useful criteria, but it is difficult to say which is better, depending on the types of problems.

Table 4.3 AIC as the goodness of fit.

Order	$p=1$	$p=2$	$p=3$
$k=p+1$	2	3	4
RSS	36.3485	0.0045	0.0080
AICc	13.64	−29.07	−24.19

Example 21

Now let revisit an earlier example (Example 19) in Section 4.2.5 using the AIC. We know that $n=5$ as there are five data points. Using the AIC criterion (4.87), we have (for $p=1$ and $k=2$)

$$\mathrm{AIC} = 2 \times 2 + 5 \ln(36.3485/5) = 13.64. \tag{4.92}$$

Similarly, we have

$$\mathrm{AIC} = -29.07, \quad -24.19 \tag{4.93}$$

for $p=2,3$, respectively. Among these three values, $p=2$ has the lowest AIC. Since the value starts to increase for $p=3$, we can conclude that the best degree of fit is $p=2$ with $k=3$ parameters. The results of AIC values are summarized in Table 4.3.

Though $p=4$ can have a 4th-order polynomial fit, the leading coefficient 0.01 (see Section 4.2.5) is too small, compared with other coefficients. This case should not be considered as it is an indication of overfitting. Ideally, a properly scaled and properly fit polynomial should have coefficients of $O(1)$. This point becomes clearer when we discuss regularization in the next section.

4.5 Regularization and Lasso method

Regularization is another approach to deal with overfitting. If we use a general model

$$y = f(\mathbf{Z}) + \epsilon, \tag{4.94}$$

where the errors ϵ obey a zero-mean normal distribution with variance σ^2, that is, $\epsilon \sim N(0, \sigma^2)$.

For multivariate cases with p components, we have

$$\mathbf{Z} = (Z_1, Z_2, \ldots, Z_p)^T. \tag{4.95}$$

In the case of linear regression, we have

$$y = \beta_0 + \beta_1 Z_1 + \beta_2 Z_2 + \cdots + \beta_p Z_p = \mathbf{Z}^T \boldsymbol{\beta} + \beta_0, \tag{4.96}$$

where β_0 is the bias, and $\boldsymbol{\beta} = (\beta_1, \beta_2, \ldots, \beta_p)^T$ is the coefficient vector.

The standard method of least squares is minimizing the residual sum of squares (RSS), that is,

$$\text{minimize} \quad \sum_{i=1}^{n} (y_i - \mathbf{Z}_i^T \boldsymbol{\beta} - \beta_0)^2. \tag{4.97}$$

The Ridge regression uses a penalized RSS of the form

$$\text{minimize} \quad \sum_{i=1}^{n} (y_i - \mathbf{Z}_i^T \boldsymbol{\beta} - \beta_0)^2 + \lambda \sum_{j=1}^{p} \beta_j^2, \tag{4.98}$$

where λ is the penalty coefficient that controls the amount of regularization. This formula can be written compactly as

$$\text{minimize} \quad ||y - \beta_0 - \mathbf{Z}^T \boldsymbol{\beta}||_2^2 + \lambda ||\boldsymbol{\beta}||_2^2, \tag{4.99}$$

where $||\cdot||_2$ is the L_2-norm, and this regularization term is based on the Tikhonov regularization on the parameter/coefficient vector. Here the bias β_0 is not part of the penalty or regularization term. One reason is that we can always pre-process the data y_i (for example, by subtracting their mean value) so that the bias β_0 becomes zero.

In the case of $\lambda = 0$, there is no penalty, which degenerates to the standard least squares. In fact, λ is a hyperparameter, which needs to be tuned.

The Lasso method uses the L_1-norm in the regularization term [139]

$$\text{minimize} \quad ||y - \beta_0 - \mathbf{Z}^T \boldsymbol{\beta}||_2^2 + \lambda ||\boldsymbol{\beta}||_1, \tag{4.100}$$

which is equivalent to the following minimization problem:

$$\text{minimize} \quad \sum_{i=1}^{n} (y_i - \beta_0 - \mathbf{Z}_i^T \boldsymbol{\beta})^2 \tag{4.101}$$

subject to

$$||\boldsymbol{\beta}||_1 = |\beta_1| + |\beta_2| + \cdots + |\beta_p| \le t, \tag{4.102}$$

where $t > 0$ is a predefined hyperparameter. Here β_0 is a bias, which is not penalized in the Lasso formulation.

A hybrid method is the elastic net regularization or regression [165], which combines the Ridge and Lasso methods into a hybrid as

$$\text{minimize} \quad ||y - \beta_0 - \mathbf{Z}^T \boldsymbol{\beta}||_2^2 + \lambda_1 ||\boldsymbol{\beta}||_1 + \lambda_2 ||\boldsymbol{b}||_2^2, \tag{4.103}$$

where both the L_1-norm and L_2-norm are used with two regularization hyperparameters λ_1 and λ_2.

4.6 Notes on software

There are a wide spectrum of software packages for regression and data mining. It is not possible to review even a good fraction of these packages. However, we will focus on a few popular tools and programming languages that most university courses are using.

- Matlab: Matlab has some well-tested curve-fitting tools `fit` and `polyfit`, the nonlinear least squares `lsqnonlin` (including the Levenberg–Marquardt method), generalized least squares `fitnlm` and `lsqcurvefit`, the Lasso method `lassglm`, the principal component analysis `pca`, and many others.
- Octave: Octave has linear least squares `lsqlin`, exponential fit `expfit`, the Levenberg–Marquardt nonlinear regression `leqsqr`, the polynomial fit `polyfit-inf`, and the nonlinear least squares `lsqnonlin`. In addition, it also has the logistic regression `logistic-regress()`.
- R: R has many functions for processing and visualizing data, including the linear regression `lm()`, the least squares fit `lsfit`, and many other statistical functionalities.
- Python: Python has at least two modules for regression `statsmodels` and `Scikit-learn`. The module Scikit learn `sklearn` is mainly for data mining and machine learning, which can do k-means clustering, basic clustering, and various statistical analysis.
- Mathematica: Mathematica can also do data mining with interface with Excel and databases. For example, it has `bestfit` and `FindFit`, the nonlinear regression `Non-linearRegress`, and the least squares `LeastSqures`. It can also do clustering via `FindClusters` with various distance metrics.

There are so many available software packages, and it is difficult to choose an appropriate one for beginners. The choice can largely depend on the availability of a package or programming language, ease of usage, and the personal expertise. For example, wikipedia has some extensive lists of

- optimization software,[1]
- data mining and machine learning,[2]
- deep learning software. [3]

We refer the interested readers to them for more detail. Here we have highlighted only a few software packages, which are either open sources, free or commercial, or easy to learn and use.

[1] https://en.wikipedia.org/wiki/List_of_optimization_software.

[2] https://en.wikipedia.org/wiki/Category:Data_mining_and_machine_learning_software.

[3] https://en.wikipedia.org/wiki/Comparison_of_deep_learning_software.

Logistic regression, PCA, LDA, and ICA

5

Contents

Following the data-fitting models and regression in the previous chapter, we now introduce logistic regression and other models for data analysis.

5.1 Logistic regression

In the previous analysis, all the dependent variable values y_i are continuous. In some applications such as biometrics and classifications, the dependent variable is just discrete or simply a binary categorical variable, taking two values 1 (yes) and 0 (no). In this case, a more appropriate regression tool is the logistic regression developed by David Cox in 1958. The logistic regression is a form of supervised learning [1,3,38].

Before we introduce the formulation of logistic regression, let us define two functions: the logistic function S and logit function. A logistic function (see Fig. 5.1), also called the sigmoid function, is defined as

$$S(x) = \frac{1}{1+e^{-x}} = \frac{e^x}{1+e^x}, \quad x \in \mathbb{R}, \tag{5.1}$$

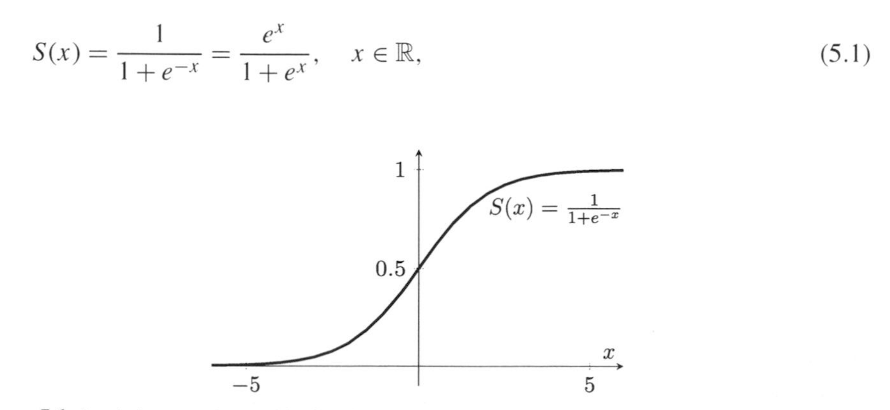

Figure 5.1 Logistic regression and its function.

Introduction to Algorithms for Data Mining and Machine Learning. https://doi.org/10.1016/B978-0-12-817216-2.00012-0

which can be written as

$$S(x) = \frac{1}{2}\Big[1 + \tanh\frac{x}{2}\Big], \quad \tanh x = \frac{e^x - x^{-x}}{e^x + e^{-x}}. \tag{5.2}$$

It is easy to see that $S \to +1$ as $x \to +\infty$, whereas $S \to 0$ as $x \to -\infty$. Thus the range of S is $(0, 1)$.

This function has an interesting property for differentiation. From the differentiation rules we have

$$\begin{aligned} S'(x) &= \Big[\frac{1}{1+e^{-x}}\Big]' = \frac{-1}{(1+e^{-x})^2}(-e^{-x}) = \frac{(1+e^{-x})-1}{(1+e^{-x})^2} \\ &= \frac{1}{(1+e^{-x})} - \frac{1}{(1+e^{-x})^2} = \frac{1}{1+e^{-x}}\Big[1 - \frac{1}{(1+e^{-x})}\Big] \\ &= S(x)[1 - S(x)], \end{aligned} \tag{5.3}$$

which means that its first derivative can be obtained by multiplication. This property can be very useful for finding the weights of artificial neural networks and machine learning to be introduced in Chapter 8.

To get the inverse of the logistic function, we can rewrite (5.1) as

$$S(1 + e^{-x}) = S + Se^{-x} = 1, \tag{5.4}$$

which gives

$$e^{-x} = \frac{1-S}{S}, \tag{5.5}$$

or

$$e^x = \frac{S}{1-S}. \tag{5.6}$$

Taking the natural logarithm, we have

$$x = \ln\frac{S}{1-S}, \tag{5.7}$$

which is the well-known logit function in probability and statistics. In fact, the logit function can be defined as

$$\text{logit}\,(P) = \log\frac{P}{1-P} = \ln\frac{P}{1-P}, \tag{5.8}$$

which is valid for $0 < P < 1$.

The simple logistic regression with one independent variable x and a binary dependent variable $y \in \{0, 1\}$ with data points (x_i, y_i) $(i = 1, 2, \ldots, n)$ tries to fit a model of logistic probability

$$P = \frac{1}{1 + e^{a+bx}}, \tag{5.9}$$

which can be written by using the logit function as

$$\ln \frac{P}{1-P} = a + bx, \tag{5.10}$$

and thus it becomes a linear model in terms of the logit of probability P. In fact, the odds can be calculated from probability by

$$O_d(\text{odd}) = \frac{P}{1-P} \tag{5.11}$$

or

$$P = \frac{O_d}{1+O_d}, \tag{5.12}$$

which means that the logistic regression can be considered as a linear model of log(odds) to x.

One naive way to solve the regression model (5.9) is to convert it to a nonlinear least squares, and we have

$$\text{minimize} \sum_{i=1}^{n} \Big[y_i - \frac{1}{1+e^{a+bx_i}} \Big]^2, \tag{5.13}$$

so as to find the optimal a and b. This is equivalent to fitting the logistic model to the data directly so as to minimize the overall fitting errors. This can give a solution to the parameters, but this is not the true logistic regression.

However, a more rigorous mathematical model exists for the binary outcomes y_i and the objective is to maximize the log-likelihood of the model with the right parameters to explain the data. Thus, for a given data set (x_i, y_i) with binary values of $y_i \in \{0, 1\}$, the proper binary logistic regression is to maximize the log-likelihood function, that is,

$$\text{maximize} \ \log(L) = \sum_{i=1}^{n} \Big[y_i \ln P_i + (1-y_i) \ln(1-P_i) \Big], \tag{5.14}$$

where

$$P_i = \frac{1}{1+e^{a+bx_i}} \quad (i = 1, 2, \ldots, n). \tag{5.15}$$

This is based on the theory of the maximum likelihood probability. Since $y_i = 1$ (yes or true) or 0 (no or false), the random variable Y for generating y_i should obey a Bernoulli distribution for probability P_i, that is,

$$B_P(Y = y_i) = P_i^{y_i} (1-P_i)^{1-y_i}, \tag{5.16}$$

so the joint probability of all data gives the likelihood function

$$L = \prod_{i=1}^{n} P(x_i)^{y_i}(1 - P(x_i))^{1-y_i}, \tag{5.17}$$

whose logarithm is given in (5.14). The maximization of L is equivalent to the maximization of $\log L$. Therefore, the binary logistic regression is to fit the data so that the log-likelihood is maximized.

In principle, we can solve the optimization problem (5.14) by Newton's method or any other optimization techniques. Let us use an example to explain the procedure in detail.

Example 22

To fit a binary logistic regression using

$$x: 0.1, \quad 0.5, \quad 1.0, \quad 1.5, \quad 2.0, \quad 2.5,$$
$$y: 0, \quad 0, \quad 1, \quad 1, \quad 1, \quad 0,$$

we can use the following form:

$$P_i = \frac{1}{1 + \exp(a + bx_i)} \quad (i = 1, 2, \ldots, 6), \tag{5.18}$$

starting with initial values $a = 1$ and $b = 1$.

Then we can calculate P_i with $a = 1$ and $b = 1$, and we have

$$P_i = \begin{pmatrix} 0.2497 & 0.1824 & 0.1192 & 0.0759 & 0.0474 & 0.0293 \end{pmatrix}.$$

The log-likelihood for each datapoint can be calculated by

$$L_i = y_i \ln P_i + (1 - y_i)\ln(1 - P_i),$$

and we have

$$L_i = \begin{pmatrix} -0.2873 & -0.2014 & -2.1269 & -2.5789 & -3.0486 & -0.0298 \end{pmatrix}$$

with the log-likelihood objective

$$\sum_{i=1}^{6} L_i = -8.2729.$$

If we try to modify the values of a and b by Newton's method, then after about 20 iterations, we should have

$$a = 0.8982, \quad b = -0.7099, \quad L_{\max} = -3.9162.$$

This means that the logistic regression model is

$$P = \frac{1}{1 + \exp(0.8982 - 0.7099x)}.$$

This logistic regression has only one independent variable. In the case of multiple independent variables $\tilde{x}_1, \tilde{x}_2, \ldots, \tilde{x}_m$, we can extend the model as

$$y = \frac{1}{1 + e^{w_0 + w_1\tilde{x}_1 + w_2\tilde{x}_2 + \ldots + w_m\tilde{x}_m}}. \tag{5.19}$$

Here we use $\tilde{x}$ to highlight its variations. To write them compactly, let us define

$$\tilde{\boldsymbol{x}} = [1,\ \tilde{x}_1,\ \tilde{x}_2,\ \ldots, \tilde{x}_m]^T \tag{5.20}$$

and

$$\boldsymbol{w} = [w_0,\ w_1,\ w_2,\ \ldots, w_m]^T, \tag{5.21}$$

where we have used 1 as a variable $\tilde{x}_0$ so as to eliminate the need to write w_0 everywhere in the formulas. Thus the logistic model becomes

$$P = \frac{1}{1 + \exp(\boldsymbol{w}^T \tilde{\boldsymbol{x}})}, \tag{5.22}$$

which is equivalent to

$$\text{logit} P = \ln \frac{P}{1 - P} = \boldsymbol{w}^T \tilde{\boldsymbol{x}}. \tag{5.23}$$

For all the data points $\tilde{\boldsymbol{x}}_i = [1, \tilde{x}_1^{(i)}, \ldots, \tilde{x}_m^{(i)}]$ with $y_i \in \{0, 1\}$ $(i = 1, 2, \ldots, n)$, we have

$$\text{maximize} \ \ \log(L) = \sum_{i=1}^{n} \Big[y_i \ln P_i + (1 - y_i) \ln(1 - P_i) \Big], \tag{5.24}$$

where $P_i = 1/[+\exp(\boldsymbol{w}^T \tilde{\boldsymbol{x}}_i)]$. The solution procedure is the same as before and can be obtained by any appropriate optimization algorithm.

Obviously, the binary logistic regression can be extended to the case with multiple categories, that is, y_i can take $K \geq 2$ different values. In this case, we have to deal with the so-called multinomial logistic regression.

Though logistic regression can work well in many applications, it does have serious limitations [120]. Obviously, it can only work for discrete dependent variables, whereas a correct identification of independent variables is a key for the model. It usually requires a sufficiently large sample size, and the sample points should be independent of each other, so that repeated observations may cause problems. In addition, it can also vulnerable to overfitting.

5.2 Softmax regression

Logistic regression is a binary classification technique with label $y_i \in \{0, 1\}$. For multiclass classification with $y_i \in \{1, 2, \ldots, K\}$, we can extend the logistic regression to the softmax regression. The labels for K different classes can be other real values, but for simplicity they can always be converted or relabeled to values from 1 to K. Softmax regression is also called multinomial logistic regression.

The softmax regression model for probability $P(y = k|\tilde{\boldsymbol{x}})$ for $k = 1, 2, \ldots, K$ takes the following form:

$$P(\tilde{\boldsymbol{x}}) = \frac{e^{\boldsymbol{w}^T \tilde{\boldsymbol{x}}}}{\sum_{j=1}^{K} e^{\boldsymbol{w}^T \tilde{\boldsymbol{x}}}}, \tag{5.25}$$

where $\boldsymbol{w} = [w_0, w_1, \ldots, w_m]^T$ are the model parameters, and w_0 is the bias. The m independent variables or attributes are written as a vector $\tilde{\boldsymbol{x}} = [1, x_1, \ldots, x_m]^T$. The denominator $\sum_1^K \exp[\boldsymbol{w}^T \tilde{\boldsymbol{x}}]$ normalizes the probabilities over all classes ensuring the sum of the probabilities to be 1.

Similar to the log-likelihood in the logistic regression, for n data samples $(\tilde{\boldsymbol{x}}_1, y_1), (\tilde{\boldsymbol{x}}_2, y_2), \ldots, (\tilde{\boldsymbol{x}}_n, y_n)$, the objective of softmax regression is to maximize the log-likelihood

$$\text{maximize } L = -\Big\{\sum_{i=1}^{n} \sum_{k=1}^{K} \mathbb{I}[y_i = k] \log\Big[\frac{e^{\boldsymbol{w}^T \tilde{\boldsymbol{x}}_i}}{\sum_{j=1}^{K} e^{\boldsymbol{w}^T \tilde{\boldsymbol{x}}_i}}\Big]\Big\}, \tag{5.26}$$

where $\mathbb{I}[\cdot]$ is the indicator function: $\mathbb{I}[y_i = k] = 1$ if $y_i = k$ is true (otherwise, $\mathbb{I} = 0$ if $y_i \neq k$). However, there is no analytical solution for this optimization problem, but we can use a gradient-based optimizer to solve it.

It is worth pointing out that softmax regression can have some redundant parameters, which means that the model is overparameterized, and thus the optimal solution will not be unique, even though the log-likelihood may be still convex. Fortunately, there is only one parameter more than necessary, and we can set the extra parameter to a fixed value such as zero. Consequently, care should be taken when using optimizers in the implementations. Most software packages have taken care of this in practice. We refer the interested readers to the more advanced literature [24,54].

5.3 Principal component analysis

For many quantities such as X_1, X_2, ..., X_p and y, it is desirable to represent the model as a hyperplane given by

$$y = \beta_0 + \beta_1 X_1 + \beta_2 X_2 + \cdots + \beta_p X_p = \boldsymbol{X}^T \boldsymbol{\beta} + \beta_0. \tag{5.27}$$

In the simplest case, we have

$$y = \beta_0 + \beta_1 x. \tag{5.28}$$

Let us first start with the simplest case; then we will extend the idea and procedure to multiple variables.

For n data points (x_i, y_i), we use the notations

$$\boldsymbol{x} = (x_1, x_2, \ldots, x_n), \tag{5.29}$$

$$\boldsymbol{y} = (y_1, y_2, \ldots, y_n), \tag{5.30}$$

so that we have

$$\boldsymbol{y} = \beta_0 + \beta_0 \boldsymbol{x}. \tag{5.31}$$

Now we can adjust the data so that their means are zero by subtracting their means $\bar{x}$ and $\bar{y}$, respectively. We have

$$\tilde{\boldsymbol{x}} = \boldsymbol{x} - \bar{x}, \quad \tilde{\boldsymbol{y}} = \boldsymbol{y} - \bar{y}, \tag{5.32}$$

and

$$\tilde{\boldsymbol{y}} + \bar{y} = \beta_0 + \beta_1(\tilde{\boldsymbol{x}} + \bar{x}), \tag{5.33}$$

which gives

$$\tilde{\boldsymbol{y}} = \beta_1 \tilde{\boldsymbol{x}}, \tag{5.34}$$

where we have used $\bar{y} = \beta_0 + \beta_1 \bar{x}$. This is essentially equivalent to removing the bias β_0 (or zero bias) because the adjusted data have zero means. In this case, the covariance can be calculated in terms of a dot product:

$$\text{cov}(\tilde{\boldsymbol{x}}, \tilde{\boldsymbol{y}}) = \frac{1}{n-1} \tilde{\boldsymbol{x}} \tilde{\boldsymbol{y}}^T. \tag{5.35}$$

In fact, the covariance matrix can be written as

$$\boldsymbol{C} = \frac{1}{n-1} \begin{pmatrix} \tilde{\boldsymbol{x}} \tilde{\boldsymbol{x}}^T & \tilde{\boldsymbol{x}} \tilde{\boldsymbol{y}}^T \\ \tilde{\boldsymbol{y}} \tilde{\boldsymbol{x}}^T & \tilde{\boldsymbol{y}} \tilde{\boldsymbol{y}}^T \end{pmatrix}. \tag{5.36}$$

Now we extend this to the case of p variables. If the mean of each variable X_i is $\bar{X}_i$ $(i = 1, 2, \ldots, p)$ with n data points, then we now use $\tilde{\boldsymbol{X}} = (\tilde{X}_1 - \bar{X}_1, \ldots, \tilde{X}_p - \bar{X}_p)^T$ to denote the adjusted data with zero means (i.e., $\tilde{X}_i = X_i - \bar{X}_i$) in the following form:

$$\tilde{\boldsymbol{X}} = \begin{pmatrix} \tilde{\boldsymbol{X}}_1 & \tilde{\boldsymbol{X}}_2 & \ldots & \tilde{\boldsymbol{X}}_p \end{pmatrix}^T = \begin{pmatrix} \tilde{X}_1^{(1)} & \tilde{X}_1^{(2)} & \ldots & \tilde{X}_1^{(n)} \\ \tilde{X}_2^{(1)} & \tilde{X}_2^{(2)} & \ldots & \tilde{X}_2^{(n)} \\ \vdots & \vdots & \ddots & \vdots \\ \tilde{X}_p^{(1)} & \tilde{X}_p^{(2)} & \ldots & \tilde{X}_p^{(n)} \end{pmatrix}, \tag{5.37}$$

$$\tilde{X}_i = X_i - \bar{X}_i. \tag{5.38}$$

Then $\tilde{X}$ is a $p \times n$ matrix, where each column is a vector for a data point. Here we used the notation $\tilde{X}_i^{(j)}$ for the jth data observations for the variable component i.

The covariance matrix can be obtained by

$$C = \frac{1}{n-1} \tilde{X} \tilde{X}^T, \tag{5.39}$$

which is a $p \times p$ symmetric matrix. If the data values are all real numbers, then C is a real symmetric matrix whose eigenvalues are all real, and the eigenvectors of two distinct eigenvalues are orthogonal to each other.

As the covariance matrix C has a size of $p \times p$, it should in general have p eigenvalues $(\lambda_1, \lambda_2, \ldots, \lambda_p)$. Their corresponding (column) eigenvectors u_1, u_2, ..., u_p should span an orthogonal matrix of size $p \times p$ by using p eigenvectors (column vectors)

$$\tilde{U} = \begin{pmatrix} \vdots & \vdots & \dots & \vdots \\ u_1 & u_2 & \dots & u_p \\ \vdots & \vdots & \dots & \vdots \end{pmatrix} \in \mathbb{R}^{p \times p}, \tag{5.40}$$

which has the properties $\tilde{U}\tilde{U}^T = \tilde{U}\tilde{U}^T = I$ (the identity matrix) and

$$\tilde{U}^{-1} = \tilde{U}^T. \tag{5.41}$$

The component associated with the principal direction (of the eigenvector) of the largest eigenvalue $\lambda* = \max\{\lambda_i\}$ is the first main component. This corresponds to the rotation of base vectors so as to align the main component to this principal direction. The transformed data can be obtained by

$$Y = \tilde{U}^T \tilde{X}. \tag{5.42}$$

It is worth pointing out that we have used column vectors here, and slightly different forms of formulas may be used if the row vectors are used, which is the case in some literature.

The principal component analysis (PCA) essentially intends to transform a set of data points (for p variables that may be correlated) into a set of $k < p$ principal components, which are a set of transformed points of linearly uncorrelated variables. It is essential to identify the first (major) component with the most variations and then to identify the next component with the second most variations. Other components can be identified in a similar manner. Thus the main information comes from the covariance matrix C and the eigenvalues (as well as the eigenvectors) of the covariance matrix.

It is worth pointing out that since the variance measures the variations, it is invariant with respect to a shift, that is, $\text{var}(X + a) = \text{var}(X)$ for $a \in \mathbb{R}$. However, it is scaled as $\text{var}(aX) = a^2 \, \text{var}(X)$.

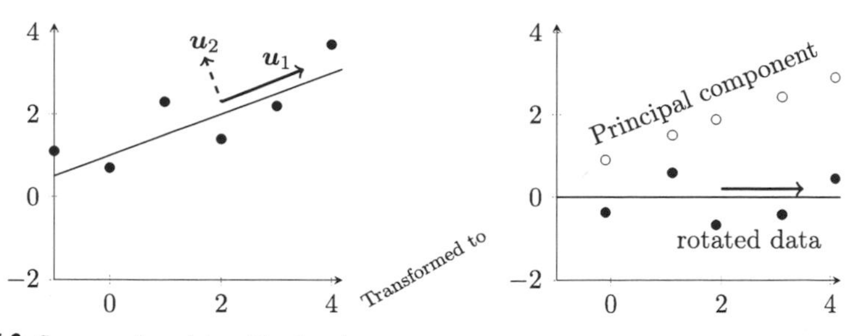

Figure 5.2 Some random data with a trend.

The main aim is to reduce redundancy in representing data. Ideally, k should be much less than p, but the first k main components should be still sufficient to represent the original data. Therefor PCA can be considered as a technique for dimensionality reduction.

The choice of the hyperparameter k can be tricky. If k is too small, then too much information is lost. If k is close to p, then almost all components are kept. So the choice of k can largely depend on the quality of the representations and information needed. A heuristic approach is based on the percentage of variance (via the eigenvalues λ_i) that can be retained in the k components. If we wish to keep μ as a percentage, then we can use

$$\text{minimize} \quad k \tag{5.43}$$

subject to

$$\sum_{i=1}^{k} \lambda_i \geq \mu \sum_{i=1}^{p} \lambda_i, \quad k \geq 1. \tag{5.44}$$

The typical value of μ can be 0.9 or higher [81,83,130].

Once $k < p$ principal components have been chosen, the data can be reconstructed easily. It is worth pointing out that some scaling and whitening are needed so as to make PCA work. In addition, PCA is a linear framework, so it cannot work for nonlinear systems. We will not delve into details of these topics; we refer the interested readers to the more advanced literature [81,132].

Let us use a simple example to demonstrate how PCA works.

Example 23

Consider the following data (see Fig. 5.2):

$$\begin{cases} x: & -1.0 & +0.0 & +1.0 & +2.0 & +3.0 & +4.0 \\ y: & +1.1 & +0.7 & +2.3 & +1.4 & +2.2 & +3.7 \end{cases}$$

First, we adjust the data by subtracting their means $\bar{x}$ and $\bar{y}$, respectively, and we have

$$\bar{x} = \frac{1}{6}\sum_{i=1}^{6} x_i = 1.5, \quad \bar{y} = \frac{1}{6}\sum_{i=1}^{6} y_i = 1.9.$$

We have $X = x - \bar{x}$ and $Y = y - \bar{y}$:

$$\begin{cases} X: & -2.5 & -1.5 & -0.5 & +0.5 & +1.5 & +2.5 \\ Y: & -0.8 & -1.2 & +0.4 & -0.5 & +0.3 & +1.8 \end{cases}$$

which have zero means. We can then calculate the covariance matrix

$$\boldsymbol{C} = \begin{pmatrix} 3.500 & 1.660 \\ 1.660 & 1.164 \end{pmatrix}.$$

Since $\text{cov}(x, y)$ is positive [so is $\text{cov}(X, Y) = \text{cov}(x, y)$], it is expected that y increases with x or vice versa.

The two eigenvalues of $\boldsymbol{C}$ are

$$\lambda_1 = 4.36, \quad \lambda_2 = 0.302.$$

Their corresponding eigenvectors are

$$\boldsymbol{u}_1 = \begin{pmatrix} 0.887 \\ 0.461 \end{pmatrix}, \quad \boldsymbol{u}_2 = \begin{pmatrix} -0.461 \\ 0.887 \end{pmatrix},$$

and these two vectors are orthogonal, that is, $\boldsymbol{u}_1^T \boldsymbol{u}_2 = \boldsymbol{u}_1 \cdot \boldsymbol{u}_2 = 0$. They span the orthogonal matrix

$$\tilde{\boldsymbol{U}} = \begin{pmatrix} 0.887 & -0.461 \\ 0461 & 0.887 \end{pmatrix}.$$

Using the adjusted data

$$\tilde{\boldsymbol{X}} = \begin{pmatrix} -2.5 & -1.5 & -0.5 & +0.5 & +1.5 & +2.5 \\ -0.8 & -1.2 & +0.4 & -0.5 & +0.3 & +1.8 \end{pmatrix},$$

we have the rotated data

$$\boldsymbol{Y} = \tilde{\boldsymbol{U}}^T \tilde{\boldsymbol{X}},$$

which gives $[0.44, -0.37, 0.59, -0.67, -0.42, 0.45]$ (see Fig. 5.2).

Since λ_1 is the largest eigenvalue, its corresponding eigenvector indicates that the x direction is the main component.

Though PCA can work well for many applications, its covariance can be very sensitive to a few large values. Thus, normalization of each dimension is needed, ideally to zero mean and unit variances. In general, PCA works under the assumption that the underlying subspace is linear. For some applications, PCA can be difficult to separate data into different classes. In this case, linear discriminant analysis can be more suitable.

5.4 Linear discriminant analysis

The linear discriminant analysis (LDA) was developed by British statistician Sir Ronald A. Fisher [3,26,47]. The main idea of Fisher's discriminant analysis is to find a transformation or a projection maximizing the separability of different classes. This can be achieved by maximizing the difference of the means of different classes while minimizing their within-class variances. The underlying assumption of LDA is that the data obey unimodal Gaussian distributions.

For n data points with m variables or dimensions, we have the data

$$\{(\boldsymbol{x}^{(1)}, y^{(1)}), (\boldsymbol{x}^{(2)}, y^{(2)}), \ldots, (\boldsymbol{x}^{(n)}, y^{(n)})\},$$

and each data point $\boldsymbol{x}^{(i)}$ (where $i = 1, 2, \ldots, n$) is an m-dimensional column vector $\boldsymbol{x}^{(i)} = (x_1^{(i)}, x_2^{(i)}, \ldots, x_m^{(i)})^T$. The output $y^{(i)}$ belongs to K different classes. In the simplest case, let us start with $K = 2$ where $y^{(i)}$ can belong to either class 1 (C_1) or class 2 (C_2) (for example, $y \in \{0, 1\}$ with $C_1 = 0$ and $C_2 = 1$). Here, n_1 data points belong to C_1, and n_2 points belong to C_2, so that $n_1 + n_2 = n$.

We can use the model

$$y = \boldsymbol{w}^T \boldsymbol{x} = w_0 + w_1 x_1 + \cdots + w_m x_m, \tag{5.45}$$

where we have used the notation similar to Eqs. (5.20) and (5.21) so as to eliminate the need to write w_0 everywhere [42].

For a two-class model, we can define their means as

$$\boldsymbol{\mu}_1 = \frac{1}{n_1} \sum_{i \in C_1} \boldsymbol{x}^{(i)}, \quad \boldsymbol{\mu}_2 = \frac{1}{n_2} \sum_{i \in C_2} \boldsymbol{x}^{(i)}, \tag{5.46}$$

which is vectors of m components. Their covariance matrices can be calculated by

$$\boldsymbol{S}_1 = \frac{1}{n_1 - 1} \sum_{i \in C_1} (\boldsymbol{x}^{(i)} - \boldsymbol{\mu}_1)(\boldsymbol{x}^{(i)} - \boldsymbol{\mu}_1)^T, \tag{5.47}$$

$$\boldsymbol{S}_2 = \frac{1}{n_2 - 1} \sum_{i \in C_2} (\boldsymbol{x}^{(i)} - \boldsymbol{\mu}_2)(\boldsymbol{x}^{(i)} - \boldsymbol{\mu}_2)^T. \tag{5.48}$$

Both are $m \times m$ matrices.

The means for the model predictions can be estimated as

$$\mu_1^{(y)} = \frac{1}{n_1} \sum_{i \in C_1} y^{(i)} = \frac{1}{n_1} \sum_{i \in C_1} \boldsymbol{w}^T \boldsymbol{x}^{(i)} = \boldsymbol{w}^T \Big(\frac{1}{n_1} \sum_{i \in C_1} \boldsymbol{x}^{(i)} \Big) = \boldsymbol{w}^T \boldsymbol{\mu}_1 \tag{5.49}$$

and

$$\mu_2^{(y)} = \frac{1}{n_2} \sum_{i \in C_2} y^{(i)} = \boldsymbol{w}^T \boldsymbol{\mu}_2. \tag{5.50}$$

In addition, the within-class variances can be calculated by

$$\begin{aligned} \bar{\sigma}_1^2 &= \frac{1}{n_1 - 1} \sum_{i \in C_1} [y^{(i)} - \mu_1^{(y)}]^2 = \frac{1}{n_1 - 1} \sum_{i \in C_1} [\boldsymbol{w}^T \boldsymbol{x}^{(i)} - \boldsymbol{w}^T \boldsymbol{\mu}_1]^2 \\ &= \boldsymbol{w}^T \Big[\frac{1}{n_1 - 1} \sum_{i \in C_1} (\boldsymbol{x}^{(i)} - \boldsymbol{\mu}_1)(\boldsymbol{x}^{(i)} - \boldsymbol{\mu}_1)^T \Big] \boldsymbol{w} = \boldsymbol{w}^T \boldsymbol{S}_1 \boldsymbol{w}. \end{aligned} \tag{5.51}$$

Similarly, we have

$$\bar{\sigma}_2^2 = \frac{1}{n_2 - 1} \sum_{i \in C_2} [y^{(i)} - \mu_2^{(y)}]^2 = \boldsymbol{w}^T \boldsymbol{S}_2 \boldsymbol{w}. \tag{5.52}$$

The main objective is to maximize their distance (or distance squared) in means

$$(\Delta \mu)^2 = |\mu_1^{(y)} - \mu_2^{(y)}|^2 = |\boldsymbol{w}^T \boldsymbol{\mu}_1 - \boldsymbol{w}^T \boldsymbol{\mu}_2|^2 = [\boldsymbol{w}^T (\boldsymbol{\mu}_1 - \boldsymbol{\mu}_2)]^2 \tag{5.53}$$

and to simultaneously minimize their within-class variances

$$\bar{\sigma}_1^2 + \bar{\sigma}_2^2 = \boldsymbol{w}^T (n_1 \boldsymbol{S}_1 + n_2 \boldsymbol{S}_2) \boldsymbol{w} = \boldsymbol{w}^T \boldsymbol{S} \boldsymbol{w}, \tag{5.54}$$

where $\boldsymbol{S} = n_1 \boldsymbol{S}_1 + n_2 \boldsymbol{S}_2$. Thus, the objective function can be written as

$$\text{maximize } f(\boldsymbol{w}) = \frac{[\boldsymbol{w}^T (\boldsymbol{\mu}_1 - \boldsymbol{\mu}_2)]^2}{\boldsymbol{w}^T \boldsymbol{S} \boldsymbol{w}}. \tag{5.55}$$

This is an optimization problem. It can be shown [129] that the optimal solution is

$$\boldsymbol{w} = \boldsymbol{S}^{-1} (\boldsymbol{\mu}_1 - \boldsymbol{\mu}_2). \tag{5.56}$$

In practice, we have to solve an eigenvalue problem to find the transformation or projection [42,129]. Let use an example to show how it works.

Example 24

Suppose we have 12 data points, where 6 belong to one class C_1, and 6 points belong to C_2, so we have $n_1 = n_2 = 6$ and $n = 12$. Their data points are

$$x_{C_1} = \left| \begin{array}{cccccc} 1 & 2 & 2 & 3 & 1.6 & 3 \\ \hline 2 & 1 & 1.5 & 2 & 1.7 & 3 \end{array} \right|, \quad x_{C_2} = \left| \begin{array}{cccccc} 5 & 6 & 7 & 8 & 9 & 7 \\ \hline 4 & 5 & 4 & 5.5 & 6.5 & 8 \end{array} \right|. \tag{5.57}$$

It is straightforward to calculate that

$$\mu_1 = \begin{pmatrix} 2.1 \\ 1.7 \end{pmatrix}, \quad \mu_2 = \begin{pmatrix} 7 \\ 5.5 \end{pmatrix}. \tag{5.58}$$

The covariances are

$$S_1 = \frac{1}{5} \begin{pmatrix} 3.1 & 2.3 \\ 2.3 & 2.8 \end{pmatrix} = \begin{pmatrix} 0.62 & 0.46 \\ 0.46 & 0.56 \end{pmatrix}, \quad S_2 = \begin{pmatrix} 2 & 1.1 \\ 1.1 & 2.4 \end{pmatrix}. \tag{5.59}$$

Thus we have

$$S = n_1 S_1 + n_2 S_2 = \begin{pmatrix} 13.1 & 7.8 \\ 7.8 & 14.8 \end{pmatrix} \tag{5.60}$$

and

$$w = S^{-1}(\mu_1 - \mu_2) = -\begin{pmatrix} 0.3223 \\ 0.0869 \end{pmatrix}, \tag{5.61}$$

which can be normalized into a unit vector as

$$w_n = \frac{w}{||w||} = \begin{pmatrix} -0.9655 \\ -0.2603 \end{pmatrix} \tag{5.62}$$

and has the same direction as

$$\begin{pmatrix} 0.9655 \\ 0.2603 \end{pmatrix}. \tag{5.63}$$

This is the direction onto which the data points should project so as to give the maximum separability of the two classes.

Both LDA and PCA can be very effective for many applications such as pattern recognition and facial recognition. In general, LDA can produce better results for pattern recognitions, though PCA can perform better for a small training data set [100].

5.5 Singular value decomposition

A real matrix $\boldsymbol{A}$ of size $m \times n$ can always be written as

$$\boldsymbol{A} = \boldsymbol{P} \boldsymbol{D} \boldsymbol{Q}^T, \tag{5.64}$$

where $\boldsymbol{P}$ is an $m \times m$ orthonormal matrix, $\boldsymbol{Q}$ is an $n \times n$ orthonormal matrix, and $\boldsymbol{D}$ is an $m \times n$ diagonal matrix. This representation is called the singular value decomposition (SVD).

A square matrix is orthogonal if each column is orthogonal to each other. A square matrix $\boldsymbol{Q}$ is orthonormal if it is orthogonal and each column is a unit vector. In other words, we have

$$\boldsymbol{Q} \boldsymbol{Q}^T = \boldsymbol{Q}^T \boldsymbol{Q} = \boldsymbol{I}, \quad \boldsymbol{Q}^{-1} = \boldsymbol{Q}^T, \tag{5.65}$$

where $\boldsymbol{I}$ is the identity matrix of the same size as $\boldsymbol{Q}$. Similarly, we have

$$\boldsymbol{P} \boldsymbol{P}^T = \boldsymbol{P}^T \boldsymbol{P} = \boldsymbol{I}, \quad \boldsymbol{P}^{-1} = \boldsymbol{P}^T. \tag{5.66}$$

In addition, the $m \times n$ diagonal matrix $\boldsymbol{D}$ is defined by $D_{ii} = d_i \geq 0$ and $D_{ij} = 0$ when $i \neq j$ for all $i = 1, \ldots, m$ and $j = 1, 2, \ldots, n$. For example, for $m = 3$ and $n = 4$, we have

$$\boldsymbol{D} = \begin{pmatrix} d_1 & 0 & 0 & 0 \\ 0 & d_2 & 0 & 0 \\ 0 & 0 & d_3 & 0 \end{pmatrix}. \tag{5.67}$$

There are rigorous algorithms to carry out SVD effectively, and we refer the interested readers to the more specialized literature [115,140].

Example 25

For example, the matrix

$$\boldsymbol{A} = \begin{pmatrix} 1 & 2 & 1 \\ 1 & 2 & 1 \end{pmatrix}$$

can be written as

$$\boldsymbol{A} = \boldsymbol{P} \boldsymbol{D} \boldsymbol{Q}^T,$$

where

$$\boldsymbol{P} = \frac{1}{\sqrt{2}} \begin{pmatrix} 1 & -1 \\ 1 & 1 \end{pmatrix}, \quad \boldsymbol{D} = \begin{pmatrix} 2\sqrt{3} & 0 & 0 \\ 0 & 0 & 0 \end{pmatrix}, \quad \boldsymbol{Q} = \frac{1}{\sqrt{6}} \begin{pmatrix} 1 & -\sqrt{3} & -\sqrt{2} \\ 2 & 0 & \sqrt{2} \\ 1 & \sqrt{3} & -\sqrt{2} \end{pmatrix}.$$

It is straightforward to check that $\boldsymbol{P} \boldsymbol{P}^T = \boldsymbol{P}^T \boldsymbol{P} = \boldsymbol{I}$ and $\boldsymbol{Q}^T \boldsymbol{Q} = \boldsymbol{Q} \boldsymbol{Q}^T = \boldsymbol{I}$.

The PCA we discussed earlier is based on the second-order statistics such as the variances and tries to find the directions with the most variations in the data. This can be achieved by finding the largest eigenvalue of the covariance matrix. The data points are then projected onto that direction. Then each succeeding step aims to find a subsequent direction that has the most variance. The projections are done in terms of Eq. (5.42).

If we can rewrite $\tilde{\boldsymbol{X}}$ using the SVD

$$\tilde{\boldsymbol{X}} = \boldsymbol{P}\boldsymbol{D}\boldsymbol{Q}^T, \tag{5.68}$$

we have the covariance matrix

$$\begin{aligned} \boldsymbol{C} &= \frac{1}{n-1}\tilde{\boldsymbol{X}}\tilde{\boldsymbol{X}}^T = \frac{1}{n-1}(\boldsymbol{P}\boldsymbol{D}\boldsymbol{Q}^T)(\boldsymbol{P}\boldsymbol{D}\boldsymbol{Q}^T)^T \\ &= \frac{1}{n-1}\boldsymbol{P}\boldsymbol{D}\boldsymbol{Q}^T\boldsymbol{Q}\boldsymbol{D}^T\boldsymbol{P}^T = \frac{1}{n-1}\boldsymbol{P}\boldsymbol{D}^2\boldsymbol{P}^T = \boldsymbol{P}\boldsymbol{\Lambda}\boldsymbol{P}^T, \end{aligned} \tag{5.69}$$

where we have used $\boldsymbol{Q}^T\boldsymbol{Q} = \boldsymbol{I}$. Here $\boldsymbol{\Lambda} = \boldsymbol{D}^2/(n-1)$ is a diagonal matrix, where the diagonal values are the eigenvalues of the corresponding eigenvectors (or the columns of $\boldsymbol{P}$).

Thus the data projection in PCA is equivalent to

$$\boldsymbol{Y} = \tilde{\boldsymbol{U}}^T\tilde{\boldsymbol{X}} = (\tilde{\boldsymbol{U}}^T\boldsymbol{P})\boldsymbol{D}\boldsymbol{Q}^T. \tag{5.70}$$

This means that PCA can be viewed as a particular class of SVD. Both PCA and SVD can be considered as a dimensionality reduction technique. A related discussion of the relationship between PCA and SVD can be found in [130,115].

5.6 Independent component analysis

The directions obtained in PCA are orthogonal, and the information used in PCA is mainly means and variances (up to the second order). It would be advantageous to use high-order statistical information for some data sets, which leads to the independent component analysis (ICA) [79,80,85,31].

The main idea of ICA is that the n observations x_i $(i = 1, 2, \ldots, n)$ are a linear mixture of n independent components s_j $(j = 1, 2, \ldots, n)$, that is,

$$x_i = a_{i1}s_1 + a_{i2}s_2 + \cdots + s_{in}s_n = \sum_{j=1}^{n} a_{ij}s_j. \tag{5.71}$$

These independent components can be considered as source signals or as hidden random variables, and thus are called latent variables or hidden features because they cannot be directly observed [80]. The equation can be written in the matrix form

$$\boldsymbol{x} = \boldsymbol{A}\boldsymbol{s}, \tag{5.72}$$

where

$$\boldsymbol{x} = [x_1, x_2, \ldots, x_n]^T, \quad \boldsymbol{A} = [a_{ij}], \quad \boldsymbol{s} = [s_1, s_2, \ldots, s_n]^T. \tag{5.73}$$

The main task is to estimate both the mixing matrix $\boldsymbol{A}$ and the unknown components $\boldsymbol{s}$. For the first, it seems that this is an impossible task because we only know the product of two unknown quantities. The ingenuity of ICA is that it uses certain structure and transformations so as to make the estimation possible [31,79].

The main assumptions for s_j $(j = 1, 2, \ldots, n)$ are that they must be mutually statistically independent and must be non-Gaussian. Otherwise, the ICA will not work. If we somehow can find an approximation (or unmixing matrix) $\boldsymbol{W}$ as an estimate to the inverse of $\boldsymbol{A}$ (or $\boldsymbol{W} \approx \boldsymbol{A}^{-1}$), then we can calculate the independent components by

$$\boldsymbol{s} = \boldsymbol{W}\boldsymbol{x}. \tag{5.74}$$

This step is equivalent to the blind source separation (BBS) in signal processing [85, 79].

For any real symmetric matrix $\boldsymbol{C}$, we have its eigenvalue problem as

$$\boldsymbol{C}\boldsymbol{u} = \lambda \boldsymbol{u}. \tag{5.75}$$

As the matrix here is real and symmetric, its eigenvectors should be orthogonal for distinct eigenvalues. Thus we can use eigenvectors to form an orthogonal matrix $\boldsymbol{U}$ whose columns are eigenvectors. We have

$$\boldsymbol{C}\boldsymbol{U} = \boldsymbol{U}\boldsymbol{\Lambda}, \tag{5.76}$$

where Λ is a diagonal matrix whose diagonal values are the eigenvalues λ_i of $\boldsymbol{C}$ with its corresponding eigenvector in ith column of $\boldsymbol{U}$. Thus we have

$$\boldsymbol{C} = \boldsymbol{U}\boldsymbol{D}\boldsymbol{U}^{-1}. \tag{5.77}$$

Since $\boldsymbol{U}$ is an orthogonal matrix, we have $\boldsymbol{U}^T = \boldsymbol{U}^{-1}$. We now have

$$\boldsymbol{C} = \boldsymbol{U}\boldsymbol{D}\boldsymbol{U}^T, \tag{5.78}$$

which is the well-known eigenvalue decomposition (EVD).

In order for the ICA to work properly, we have to center the observations by subtracting the mean or expectation $\mu_i = E[x_i]$ from x_i, that is, $x_i - \mu_i$, which means that the covariance matrix $\boldsymbol{C} = E[\boldsymbol{x}\boldsymbol{x}^T]$ is symmetric and real, and thus we can write it

$$\boldsymbol{C} = E[\boldsymbol{x}\boldsymbol{x}^T] = \boldsymbol{U}\boldsymbol{D}\boldsymbol{U}^T. \tag{5.79}$$

However, we can scale $\boldsymbol{x}$ so that its variance becomes the identity matrix $\boldsymbol{I}$. By defining a scaled vector

$$\tilde{\boldsymbol{x}} = (\boldsymbol{U}\boldsymbol{D}^{-1/2}\boldsymbol{U}^T)\boldsymbol{x} \tag{5.80}$$

with $D^{-1/2} = \text{diag}\{d_1^{-1/2}, \ldots, d_n^{-1/2}\}$, we have its covariance matrix

$$\tilde{C} = E[\tilde{x}\tilde{x}^T] = I. \tag{5.81}$$

This step of rescaling is called whitening. This is in fact equivalent to normalizing the variances after rotating the centered observations $\bar{x}_i$ to align with the eigenvectors of the covariance matrix so that they are along the rotated Cartesian basis vectors [131]. As a result, all linear dependencies in the data have been removed.

The whitening step essentially transforms A to $\tilde{A}$:

$$\tilde{A} = \left(U D^{-1/2} U^T\right) A, \quad \tilde{x} = \tilde{A} s, \tag{5.82}$$

so that

$$E[\tilde{x}\tilde{x}^T] = E[\tilde{A}s(\tilde{A}s)^T = \tilde{A}E[ss^T]\tilde{A}^T = \tilde{A}\tilde{A}^T = I, \tag{5.83}$$

where we have assumed that sources s_j are also whitened so that $E[ss^T] = 1$ [131,79]. The main advantage of these steps is that the number of degrees of freedom is now reduced to $n(n-1)/2$ due to the unique structure of orthogonal matrices.

Now to find A (or $\tilde{A}$) or the estimate inverse W is to find a proper rotation matrix U. Alternatively, we can view this as an optimization problem to maximize the statistical independence of s_j, which is equivalent to the minimization of the multiinformation $I(s)$ of the unknown distribution $f(s)$,

$$I(s) - \int f(s) \log_2 \left[\frac{f(s)}{\Pi_i f(s_i)}\right] ds, \tag{5.84}$$

which is always nonnegative with the minimum at $I(s) = 0$ if all variables are statistically independent when $f(s) = \Pi_i f(s_i)$ [130]. This can be approximated by

$$I(s) \approx \sum_i H(\tilde{A}s_i) = \sum_i H(\tilde{x}_i), \tag{5.85}$$

where

$$H(x) = -\int f(x) \log_2 f(x) dx \tag{5.86}$$

is the entropy of the probability distribution $f(x)$. This optimization problem is subject to the constraint $WW^T = I$, which can be solved effectively by algorithms such as the FastICA and maximum likelihood [79,80].

Both PCA and ICA have been used in a diverse range of applications. They share some similarities, but they do have some important differences. For example, the correlation in the data is gradually removed in PCA, and some components are more dominant than others. In contrast, both higher-order dependence and correlations are removed in ICA. In addition, ICA mainly assumes that the underlying distributions must not be Gaussian, and the vectors are not necessarily orthogonal in ICA, unlike the orthogonality in PCA. Furthermore, as the result of its whitening step, all components are essentially equally important in the framework of ICA. These characteristics mean that ICA can have certain advantages over PCA in some applications [39].

5.7 Notes on software

Almost all major software packages in statistics, data mining, and machine learning have implemented the algorithms we introduced in this chapter. For example, R has `pca` and `fastICA` packages. Python has `pca`, `ica` from `sklearn`. Matlab has all the functionalities such as `pca`, `svd`, and `mnrfit` for multinomial logistic regression.

Data mining techniques

6

Contents

The evolution of the Internet and social media has resulted in the huge increase of data in terms of both volumes and complexity. In fact, "big data" has become a buzzword nowadays, and the so-called big data science is becoming an important area.

Data mining has expanded beyond the traditional data modeling techniques such as statistical models and regression methods. Data mining now also includes clustering and classifications, feature selection, and feature extraction and machine learning techniques such as decision tree methods, hidden Markov models, artificial neural networks, and support vector machines. To introduce these methods systematically even by taking a whole book [144–146] it is not possible to cover even a good fraction of these methods in a book chapter. Therefore, we focus on some of the most widely used methods.

Loosely speaking, for a multidimensional data sample with D different attributes or features in a D-dimensional space $\boldsymbol{x} = (x_1, x_2, \ldots, x_D)^T$, if the dependent variable y is a continuous variable, then the problem becomes a regression problem. If y only takes some discrete values (class labels) such as 0 and 1 (or some integer values), then it becomes a classification problem. In the special case of no class labels at all, such a classification problem becomes a clustering problem [96,138,163].

Clustering and classification methods are rather rich with a wide spectrum of methods. We introduce the basic k-mean method for clustering and support vector machine

Introduction to Algorithms for Data Mining and Machine Learning. https://doi.org/10.1016/B978-0-12-817216-2.00013-2

for classification. Artificial neural networks (ANN) are a class of methods with different variations and variants, and ANN can have many applications in a diverse range of areas, including clustering, classification, machine learning, computational intelligence, feature extraction and selection, and others. We will introduce ANN in Chapter 8.

6.1 Introduction

Before we introduce some data mining techniques, let us briefly discuss the types of data and distance measures.

6.1.1 *Types of data*

There are many different types of data, and there are different ways of classifying such data, depending on the emphasis on certain features such as structures, media, database, source, time, or dimensionality. Here we will loosely divide all data into three categories: structured, unstructured, and mixed.

Structure data sets have fixed structures. Many data from science and engineering are well structured, including time series signals, images, astronomical surveys, weather records, and databases. Unstructured data can be of any form without a specific structural layout. Data related to social media, world-wide webs, business transactions, and others are typically unstructured. In addition, data can be of mixed type, where some parts are well structured, whereas the other parts do not. For example, emails can mixed with text and attached data files, images, and videos. Many data and multimedia repositories can be mixtures of different data types.

The main aim of data mining is to make sense of data by processing, analyzing, and categorizing the data using various data mining techniques such as classification, clustering, feature selection, and others. Ultimately, the purpose is to understand the data with great insight and then to be able to make predictions for new data and unknown data [62].

6.1.2 *Distance metric*

It is worth pointing out that distance metrics are also very important. Even with the most efficient methods, if the metric measure is not defined properly, then the results may be incorrect or meaningless. Most clustering methods use the Euclidean distance and Jaccard similarity, though other distances such as the edit distance and Hamming distance are also widely used.

Briefly speaking, the distance $d(\boldsymbol{x}, \boldsymbol{y})$ in the D-dimensional space between two data points $\boldsymbol{x} = (x_1, x_2, \ldots, x_D)^T$ and $\boldsymbol{y} = (y_1, y_2, \ldots, y_D)^T$ is the L_p-norm given by

$$d(\boldsymbol{x}, \boldsymbol{y}) = ||\boldsymbol{x} - \boldsymbol{y}||_p = \Big(\sum_i^D |x_i - y_i|^p\Big)^{1/p}, \tag{6.1}$$

which is also called the Minkowski distance in the literature. In the case of $p = 2$, it becomes the standard Euclidean or Cartesian distance.

The Manhattan distance is defined by

$$D_m(\boldsymbol{x}, \boldsymbol{y}) = \sum_{i=1}^{D} |x_i - y_i|, \tag{6.2}$$

which is its L_1-norm.

Jaccard's similarity index of two sets U and V is defined as

$$J(U, V) = |U \cap V|/|U \cup V|, \tag{6.3}$$

which leads to $0 \le J(U, V) \le 1$, and the Jaccard distance is defined as

$$d_J(U, V) = 1 - J(U, V). \tag{6.4}$$

The edit distance between two strings U and V is the smallest number of insertions and deletions of single characters that will convert U to V. For example, $U =$ "abcde" and $V =$ "ackdeg", the edit distance is $d(U, V) = 3$. By deleting b, inserting k after c, and then inserting g after e, the string U can be converted to V.

On the other hand, the Hamming distance is the number of components by which two vectors/strings differ. For example, the Hamming distance between 10101 and 11110 is 3. Obviously, other distance metrics are also used in the literature.

6.2 Hierarchy clustering

For a given set of n observations, the aim is to divide them into some clusters (say, k different clusters) so as to minimize certain clustering measures or objectives. There are many key issues here. Firstly, we usually do not know how many clusters the data may intrinsically have. Secondly, the data sets can be massive ($n \gg 1$, for example, $n = O(10^9)$ or even $n = O(10^{18})$). Thirdly, the data may not be clean enough (often with useless information and/or noisy data). Finally, the data can be incomplete, and thus may lack sufficient information needed for correct clustering. Obviously, there are other issues, too, such as time factors, unstructured data, and distance metrics.

Hierarchy clustering usually works well for small datasets. It starts with every point in its own cluster (that is, $k = n$ for n data points), followed by a simple iterative procedure (see Algorithm 4).

Algorithm 4 Hierarchy clustering algorithm.

1: Each point belongs to its own cluster $k = n$
2: **for** all data points **do**
3: Choose two nearest clusters to merge into one cluster
4: Update $k \leftarrow k - 1$
5: Repeat until the metric measure goes up
6: **end for**

This iterative procedure can lead to one big cluster $k = 1$ in the end. But it does result in a complex decision tree, which provides an informative summary and some insight into the structures and relationship within the data. However, if a distance metric such as the Euclidean metric is defined properly, then the metric will start to decrease at the initial stage when two clusters are merged. In the final stage, this metric usually starts to increase, which is an indication to stop, and the number of the clusters can be the true number of clusters. However, this is not straightforward in practice, and there may not exist any unique k value at all.

In fact, k is a hyperparameter, which needs some tuning and parametric studies in practice.

In the case where Euclidean distance measures are used, the distance between centroids is defined as the cluster distance. The complexity of this algorithm is $O(n^3)$, where n is the number of points. For $n = 10^9$, this can lead to $O(10^{27})$ floating-point operations, which is quite computationally expensive.

6.3 *k*-Nearest-neighbor algorithm

The k-nearest-neighbor (kNN) algorithm is a voting algorithm as it uses k sample points in terms of a given distance measure to determine the class of the data point under consideration. The main steps of a kNN algorithm can be summarized as Algorithm 5.

Algorithm 5 Nearest-neighbor algorithm.

1: Define k
2: **while** (stopping criterion is not met) **do**
3: Compute distances from other data points to point i
4: Sort the computed distances
5: Select the k points with smallest distances
6: Assign the test point to the class by the simple majority
7: Return the class
8: **end while**

The essential idea of this algorithm is easy to understand, and the algorithm is relatively simple to implement. The classification is based on local information and

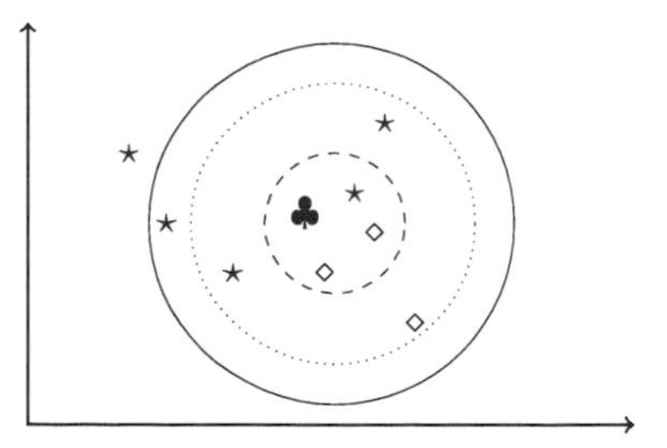

Figure 6.1 The schematic representation of kNN.

distances, and thus the decision boundaries can be rather flexible to deal with irregular complex shapes.

The kNN algorithm is quite widely used due to its simplicity, and it is yet sufficiently effective in many applications. Many online systems use it for recommending products and movies. In addition, many text and image classification systems use it. However, as the computation and sorting of distances iteratively can be computationally expensive, thus it is not suitable for big data sets.

The main idea of a simple classification problem is shown in Fig. 6.1 where there are two classes, stars ($\star$) and diamonds ($\diamond$). The task is to determine to what class $\clubsuit$ should belong? If we use $k = 3$ (the smaller dashed circle), then the unknown $\clubsuit$ should belong to $\diamond$. If we use $k = 7$ (the bigger solid circle), then this $\clubsuit$ should be classified as $\star$. However, if we use $k = 6$ (the dotted circle), it would be difficult to classify $\clubsuit$ because it leads to a tie in this case. This highlights the importance of choosing the right parameter k and its sensitivity.

To avoid a tie, k should be an odd number. If k is too small, then it may lead to overfitting and more sensitive to noise in data, whereas large k values may lead to higher bias and lower accuracy because it may include samples that are not actual neighbors. However, there is no easy way to determine k, and some iterative cross-validation may be needed. A crude guideline is $k < \sqrt{n}$ for a given set of n training data points.

Another issue is that the distances computed may be dominated by the dimensions that have largest ranges. For example, the distance between $\boldsymbol{x} = [1,\ 1000,\ 10]$ and $\boldsymbol{y} = [0.5,\ 100,\ 1]$ will be largely affected by the second dimension or attribute whose range is 1000. In this case, some rescaling or normalization is needed to make sure that all ranges are comparable, typically in the range of [0,1].

There are quite a few variants of kNN, such as kNN regression where the averages of k samples are used, fuzzy kNN, and others.

6.4 k-Means algorithm

The main aim of the k-means clustering method is to divide a set of n observations into k different clusters in such a way that each point belongs to the nearest cluster with the shortest distance to its corresponding cluster mean or centroid.

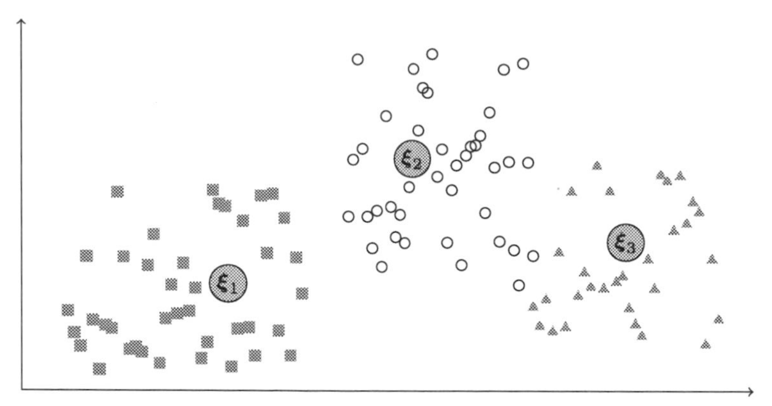

Figure 6.2 K-means algorithm and the cluster centers.

Suppose we have n observation points $\boldsymbol{x}_1, \boldsymbol{x}_2, \ldots, \boldsymbol{x}_n$ in a d-dimensional vector space. Our aim is to partition these observations into k clusters $(S_1, S_2, \ldots, S_k)$ with centroid means $(\boldsymbol{\xi}_1, \boldsymbol{\xi}_2, \ldots, \boldsymbol{\xi}_k)$ as shown in Fig. 6.2, so that the clusterwise sum of squares, also called within-cluster sum of squares, can be minimized, that is,

$$\text{minimize} \quad \sum_{j=1, \boldsymbol{x}_i \in S_i}^{k} ||\boldsymbol{x}_i - \boldsymbol{\xi}_j||^2, \tag{6.5}$$

where $1 < k \leq n$ and typically $k \ll n$.

This k-means method for dividing n points into k clusters can be summarized schematically in Algorithm 6.

Algorithm 6 k-means algorithm.

1: Choose randomly k points as the initial centroids of the k clusters
2: **for** each remaining point i **do**
3:     Assign i to the cluster with the closest centroid
4:     Update the centroid of that cluster (containing i)
5: **end for**

There are some key issues concerning this method. The choice of k points as the initial centroids is not efficient. In the worst case, the k points selected randomly can belong to the same cluster. One possible remedy is to choose k points with the largest distances from each other. This is often carried out by starting from a random point, then trying to find the second point that is as far as possible from the first point, and then trying to find the third point that is as far as possible from the previous two points. This continues until the first k points are initialized. This method is an improvement over the previous random selection method, but there is still no guarantee that the choice of these initial points will lead to the best clustering solutions. Therefore, some sort of random restart and multiple runs are needed.

On the other hand, the algorithm complexity of this clustering method is typically $O(n^{kd+1}\log(n))$, where d is the dimension of the data. Even for $n = 10^6$, $k = 3$, and $d = 2$, this becomes $O(10^{43})$, which is extremely computationally expensive. However, it is worth pointing out that such a complexity is just theoretical, and these methods can sometime work surprisingly well in practice (at least for small datasets). In the worst cases, such an algorithm complexity can become NP-hard.

For a given dataset, it is difficult to know what k should be used because k is a hyperparameter. Ideally, the initial choice of k should be sufficiently close to the actual number of intrinsic classes in the data, and then some adjustment around this initial guess can be done. But in practice, this may need either some experience or other methods to get sense of data. In addition, the clustering distances can also be used to check if k is a proper choice in many cases.

There are many other methods such as fuzzy k-means method and others. The interested readers can consult a more advance literature about data mining.

6.5 Decision trees and random forests

6.5.1 Decision tree algorithm

The decision tree algorithm is a widely used algorithm for classification, which uses attribute values to partition the decision space into smaller subspaces in an iterative manner. Its essence is a divide-and-conquer approach, starting with a root node and gradually growing to a final classification (or leaf). The decision processes can be represented graphically as a tree, though this tree is usually drawn upside down [123].

Example 26

Let us use an example about a hypothetical classification system for a university degree. Suppose a student has to take different modules and a major dissertation or thesis for a degree, and each module subject has an exam. The minimum pass marks are 40%. For each thesis, three grades (merit, pass, fail) are given. A naive system is used for this hypothetical degree programme, as shown in Fig. 6.3, to put equal weights on the exams and thesis. If a student gets all the modules with a mark above 80% with a merit of his or her thesis, a first class degree (I) will be award to this student. The final classes are first class (I), second class (II), third class (III), or pass.

This degree classification procedure can be represented by a decision tree as shown in Fig. 6.4, where the tree is drawn upside down from a root node Thesis growing to leaf nodes (circles), called terminal nodes or decision nodes, which correspond to outcome classes.

It is worth pointing out that the tree is a directed tree in the sense that decisions start at the top and go downward. A node with one incoming edge and multiple outgoing edges is called an internal node. Each possible decision is covered and represented as a branch, and a complete decision process is essentially a path or branch from the root node to a leaf.

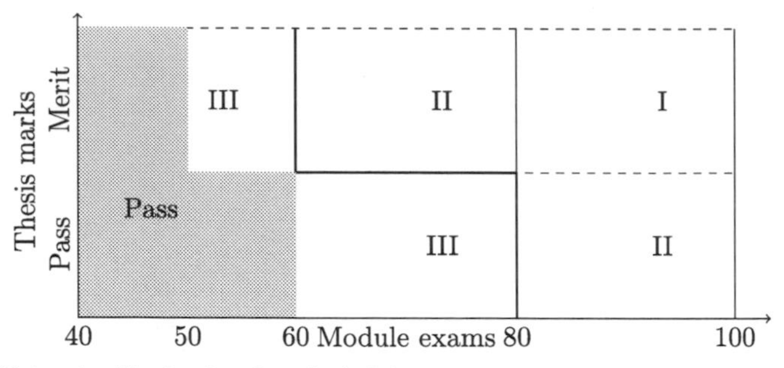

Figure 6.3 Naive classification for a hypothetical degree.

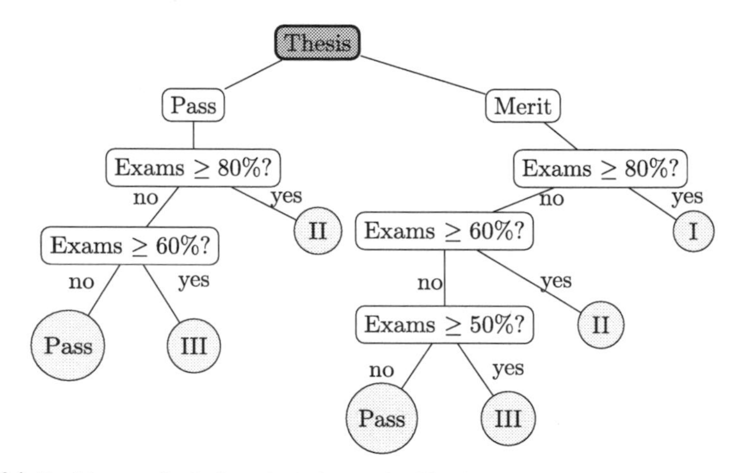

Figure 6.4 Decision tree for the hypothesis degree classification.

6.5.2 *ID3 algorithm and C4.5 classifier*

Obviously, trees can be drawn in different ways, depending on the focus and which criterion is applied first. Thus, the decision trees are not unique.

In principle, different ways of drawing the trees should lead to the same conclusions in terms of the final classification class outcomes. As the complexity increases with the number of attributes and their value ranges or categorical values, the actual tree can be highly complex, and thus decision trees may not necessarily lead to the same conclusion. This issue becomes more significant when the values of certain attributes have noise or uncertainty, or when certain attributes may have missing values.

Another issue is which attribute should be used first. Different algorithms may choose attributes using different criteria. To make an informed choice, we have to define certain measures or criteria. The popular algorithm ID3 (Iterative Dichotomiser 3), developed by Quinlan [117] in 1986, uses an information gain as a rule to select the attribute that can explain the training examples and maximize the information gain.

For a two-class classification (1 or 0) problem with n training examples, if p_1 represents the fraction of class 1 among the whole set, and $p_0 = 1 - p_1$ represents the fraction of class 0 among the whole set, then the entropy S can be defined as

$$S = -p_1 \log_2 p_1 - p_0 \log_2 p_0, \tag{6.6}$$

where base 2 logarithm is used. It is worth pointing out that here we conventionally set $0 \log_2 0 = 0$. In case of K multiple classes, this entropy can be extended to

$$S = -\sum_{i-1}^{K} p_i \log_2 p_i, \tag{6.7}$$

where p_i is the fraction or probability of samples belonging to class i. It is straightforward to check that the minimum value of S is zero, whereas the maximum value $S_{\max} = \log_2 K$ is reached when all p_i are the same, that is, $p_i = 1/K$ (a uniform distribution). In addition, in some literature, this Shannon entropy is called an impurity function or impurity measure. Another related classical impurity function is the so-called Gini index or Gini impurity

$$G_I = \sum_{i=1}^{K} p_i(1 - p_i). \tag{6.8}$$

The information gain $IG(A)$ of an attribute A with possible categorical values of x_j, where $j = 1, 2, \ldots, K$ are the categories, is defined as

$$IG(A) = S - \sum_{x_j \in \Omega(A)} f_j S_{x_j}, \tag{6.9}$$

where $\Omega(A)$ is the set of categorical values of attribute A, and f_j is the fraction of value $\boldsymbol{x}_j \in A$, which can be calculated by the ratio of the number $|A_{x_j}|$ of $x_j \in A$ to the total number of values or cardinality $|A|$. The S_{x_j} is the entropy, which has a similar formula to the entropy S, though its fractions should be based on the categorical values of that attribute.

Let us look at an example to see how these formulas work.

Example 27

For a class of a degree programme, there are 30 students. The numbers of students of different classes (I, II, III, and pass) are 12, 9, 6, 3, respectively. So we have $K = 4$ classes, and the fractions of each class are

$$p_1 = \frac{12}{30} = 0.4, \quad p_2 = \frac{9}{30} = 0.3, \quad p_3 = \frac{6}{30} = 0.2, \quad p_4(\text{pass}) = \frac{3}{30} = 0.1.$$

So their entropy is

$$S = -\sum_{i=1}^{4} p_i \log_2 p_i$$
$$= -\Big[0.4\log_2 4 + 0.3\log_2 0.3 + 0.2\log_2 0.2 + 0.1\log_2 0.1\Big] = 1.8464.$$

Among all the four classes, suppose there are 20 theses that are merit, and 10 theses that are just pass. For merits, their corresponding students' classes are 16, 2, 1, 1 for I, II, III, and pass, respectively, whereas for theses with a pass, they are 2, 2, 4, 2 for I, II, III, and pass, respectively. So the attribute "thesis" has two values "merit" and "pass" (all 30 students got a degree). So the fractions for merit and pass are 20/30 and 10/30, respectively, that is, $f_1 = 2/3$ and $f_2 = 1/3$.

For the theses with a merit, their fractions or probabilities of different classes are $p = [16, 2, 1, 1]/20 = [0.8, 0.1, 0.05, 0.05]$, and thus their entropy is

$$S_{x_1} = -[0.8\log_2 0.8 + 0.1\log_2 0.1 + 0.05\log_2 0.05 + 0.05\log_2 0.05] = 1.0219.$$

For the theses with a pass, their fractions are $p = [2, 2, 4, 2]/10 = [0.2, 0.2, 0.4, 0.2]$, and their entropy is

$$S_{x_2} = -[0.2\log_2 0.2 + 0.2\log_2 0.2 + 0.4\log_2 0.4 + 0.2\log_2 0.2] = 1.9219.$$

Finally, the information gain is

$$IG(\text{thesis}) = S - \sum_{x_j \in [\text{merit, pass}]} f_j S_{x_j}$$
$$= 1.8464 - \Big[\frac{2}{3} \times 1.0219 + \frac{1}{3} \times 1.9219\Big] = 0.5245.$$

This means that the information gain by using thesis marks as a decision criterion is 0.5245 bit.

The main idea of the ID3 algorithm is to search the decision space and form decision trees by fitting the data and maximizing the information gain as each branching or decision step [117]. The main steps of ID3 are as Algorithm 7.

Algorithm 7 The ID3 algorithm.

1: Load the data and calculate the sample entropy S
2: **for** all attributes **do**
3: Find the attribute with the maximum information gain $IG(A)$
4: Split the set into subsets by values of that best attribute
5: Create a decision tree node for that attribute with $IG(A)_{\max}$
6: Iterate over the rest of unused/unsplit attributes
7: **end for**

There are different ways to define a proper stopping criterion. Essentially, the iterations stop if no more attributes are left, or no data samples in a subset, or each entry in a subset becomes the same class, or all nodes are turned into leaf nodes.

Though the ID3 is simple and can work well, however, this method is a greedy method, and thus it tends to lead to overfit the data set, and there is no guarantee of optimality. Obviously, this method cannot handle continuous data set.

A significant improvement was the introduction of C4.5 method by Quinlan [118] in 1993, which was later ranked among the top 10 algorithms in data mining [30,147]. In contrast with ID3, C4.5 can deal with continuous data and missing data. In addition, pruning is used in C4.5 so as to reduce the depth of the trees and remove unnecessary or insignificant branches. The decision criterion uses a normalized information gain ratio. The normalization also reduces the influence of large variations of different attribute values.

For K different classes or categories, the split information (SI) of an attribute is the entropy

$$SI(A) = -\sum_{i=1}^{K} f_i \log_2 f_i, \tag{6.10}$$

where f_i is the fraction or probability of the data in category i among the whole data set. The information gain ratio (IGR) is the ratio of information gain, defined in Eq. (6.9), to the split information in Eq. (6.10), that is,

$$IGR(A) = \frac{IG(A)}{SI(A)}, \tag{6.11}$$

which is essentially the information gain, normalized by the split information.

It is worth pointing out that this normalized information gain ratio has the advantage of not preferring the attributes with nearly uniform distributions, but it does have a potential issue of dividing by a small number (or nearly zero) if splitting information is too small. In this case, an addition criterion to check the values of information gain is needed.

The main steps of the well-known C4.5 algorithm are summarized in Algorithm 8.

Algorithm 8 C4.5 algorithm.

1: Load data and initialize
2: **for** (each attribute A) **do**
3: Calculate its normalized information gain ratio $IGR(A)$
4: Find the attribute with the maximum IGR
5: Create a decision node by splitting that attribute
6: Iterate over the rest of the unsplit attributes
7: **end for**

In addition, to prevent overfitting, the C4.5 algorithm uses a so-called reduced-error pruning. The main idea of pruning is to remove the subtree or child nodes of a decision node and turn that decision node into a leaf node by giving it with the most common class among the training data samples. The removal of nodes can be according to certain criteria, and the removal of these nodes should not reduce the performance of the resulting trees, in comparison with the original larger trees [118].

The way of dealing with continuous attributes is relatively straightforward. For example, if an attribute A is continuous, then it can be simply converted into a Boolean attribute at a critical value x_* by setting true (or 1) if $A \geq x_*$ and false (or 0) if $A < x_*$. The critical value x_* can be any decision boundary that be appropriate for that attribute and decision-making.

Though C4.5 algorithm and its variants work well, the tree for each test instance constructed is a single tree. For large datasets, a natural extension is to use multiple trees simultaneously. Multiple trees are constructed from random sampling from the learning dataset with replacements from a forest of decision trees. In this case, we are dealing with the well-known random forest algorithm.

6.5.3 Random forest

The main idea of the random forest classifier to construct multiple decision trees randomly using sampling with replacement, and the final decisions will be based on a rule, often in terms of a voting system of the ensemble of the decision trees. The basic idea of random decision forests was developed by Ho [74] in 1995 and significantly extended later by Breiman [23] using the idea of bagging. Bagging can be used to fit many large trees by using a majority vote as the rule of classification. The random forest algorithm is a combination of multiple decorrelated trees with bagging. In essence, the random forest classifier is an ensemble method or ensemble-based decision-tree learning with the aim to improve accuracy with reduced variance and to potentially avoid overfitting.

For a set of D features (or dimensionality of the training dataset), we often use m features for splitting the data at each decision split, though $m = \lceil \sqrt{D} \rceil$ or $m = \lceil \log_2 D \rceil$ (rounded to the nearest integer) is usually used to determine m. For each subset of m features, the training samples are randomly selected from the original data set with replacement. Then each subset is fed into a decision tree classifier, and classification is based on the majority vote of the ensemble.

Each tree can cast a single vote for the most popular class for a given input feature vector. The basic idea of random forests are schematically represented in Fig. 6.5.

The main steps of the random forest classifier can be summarized as the steps shown in Algorithm 9.

Though random forests usually do not lead to overfitting, however, overfitting can still occur for noisy data. It is worth pointing out that there are still some free parameters (or hyperparameters) to set. For example, how many trees should be used in a forest. Typically, a few dozens to a few hundred trees are needed to estimate the statistical measures properly, and some literature uses $T = 64$ to 128 (or even 256) trees.

There are a vast array of variants, based on random forests, including the kernel random forest and various boost methods such as Adaboost. The interested readers can consult a more advanced literature on these topics.

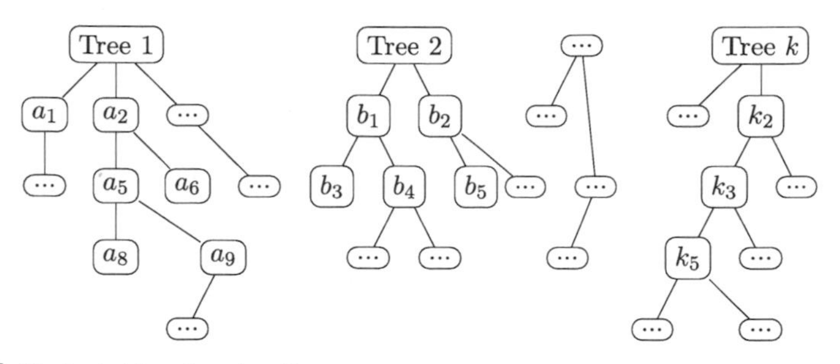

Figure 6.5 The basic idea of random forests.

Algorithm 9 Random forest algorithm.

1: Load data and initialize
2: **for** (a given number of trees) **do**
3: Select m features randomly from D features
4: Sample data to create T different trees
5: Create each tree by iteratively splitting/forming decision nodes
6: Predict any test feature by running the multiple trees
7: Classify using majority voting
8: **end for**

6.6 Bayesian classifiers

Classifications can be carried out from a probabilistic perspective, and the naive Bayesian classifier is one of the most widely used classifiers.

6.6.1 Naive Bayesian classifier

Probabilistic reasoning and inferencing can be considered as approaches based on the assumptions that decision variables obey some probability distributions and data are drawn from certain probability distributions. We can gain more insight by learning from the data and approximating the underlying distributions more accurately as more data are accumulated [127,61].

The essence of a Bayesian classifier is to estimate the probabilities of all alternative models or hypotheses, given data as evidence, and then to find the most probable classification to be assigned to each new input. The core foundation of Bayesian classifiers is the Bayes' theorem or Bayesian rule

$$p(H|S) = \frac{p(S|H)p(H)}{p(S)}, \tag{6.12}$$

which estimates the posterior probability $p(H|S)$ for a hypothesis or model, given the samples or training data set S. Here $p(S|H)$ is the likelihood probability of the data samples, given H is true, and $p(H)$ is the prior probability of hypothesis H, which somehow incorporates any background knowledge about H. If there is no prior knowledge, we can use some uniform distributions as a prior [107,108]. Furthermore, $p(S)$ can be considered as the prior probability of sample data S. Depending on the emphasis and formulations, we can seek a maximum a posterior (MAP) hypothesis

$$\text{maximize } p(H|S) \propto \text{maximize } P(S|H)P(H), \tag{6.13}$$

or we can seek a maximum likelihood (ML)

$$\text{maximize } p(S|H), \tag{6.14}$$

which is equivalent to the previous MAP if both $p(H)$ and $p(S)$ are constant.

For a classification problem with a focus on an attribute $A \doteq x$ with K different attribute values $[x_1, x_2, \ldots, x_K]$, its model function (the model hypothesis) $y = f(x)$ forms a set of finite discrete values $y_j \in \Omega$. The aim of a Bayesian classifier is to estimate the probability of y, given data x_i, so as to assign the class probability

$$\max p(y_j|x_i) = p(y_j|x_1, x_2, \ldots, x_K), \quad y_j \in \Omega, \tag{6.15}$$

which is equivalent to [from Eq. (6.12)]

$$\max \frac{p(x_1, x_2, \ldots, x_K|y_j)p(y_j)}{p(x_1, x_2, \ldots, x_K)} \propto \max p(x_1, x_2, \ldots, x_K|y_j)p(y_j). \tag{6.16}$$

However, it is not trivial to calculate the probability $p(x_1, \ldots, x_K|y_j)$, and in most cases, it is impossible to calculate at all.

A naive assumption is that all the data sample values are conditionally independent from each other, and consequently the joint probability becomes a product of individual probability. Thus we can use

$$p(x_1, x_2, ..., x_K|y_j) = \prod_{i=1}^{K} p(x_i|y_j). \tag{6.17}$$

Eq. (6.16) becomes

$$\max p(y_j) \prod_{i=1}^{K} p(x_i|y_j). \tag{6.18}$$

A probabilistic classifier using this to assign probabilities becomes a naive Bayesian classifier.

The main task now is to estimate the probability of the likelihood and the prior probability from a training data set. In some cases, especially for continuous dataset, we can assume that the samples are drawn from a Gaussian distribution. In this case, the naive Bayesian classifier becomes the Gaussian naive Bayesian classifier.

One issue is that the product of multiple probability terms can lead to very small values, or even near zero values. To avoid such a potential difficulty, Mitchell suggested to use an m-estimate of probability using

$$\frac{K_j + m_s p}{K + m_s}, \tag{6.19}$$

where K_j is the number of samples with $y = y_j$, and p is a prior estimate of the probability (which is usually taken as a uniform distribution), and m_s is the equivalent sample size. For more detail, the readers can refer to the book by Michell [105] and other specialized literature [51,108].

6.6.2 Bayesian networks

A key assumption for naive Bayesian classifiers is that all variable values are conditionally independent, given the target classification. This assumption significantly reduces the complexity of the calculations of the objective functions in terms of posterior probabilities.

However, this assumption may not be true for certain applications such as text documents and speech signals. In this case, Bayesian belief networks can be a good alternative.

Bayesian belief networks (BBN) use a set of conditionally independent probabilities, but not imposing all variable values. We have seen some basic idea of Bayesian networks in Chapter 2, and here we focus on the classification problems. In a BBN, nodes represent variables that can be continuous or discrete, and arcs represent causality relationships in terms of conditional probabilities.

For a given structure of a BBN, not every variable is observable. Unobservable variables are called hidden or latent variables. The BBN model has both observable random variables X and hidden random variables Z, which means that the likelihood function $p(S|H)$ becomes a function of X and Z, that is, $L(X, Z)$. The maximization of the likelihood $L(X, Z)$ is equivalent to the maximization of expectation of the logarithmic likelihood function. Thus the method becomes the so-called expectation-maximization (EM) method, which consists of an E step and an M step [36]. In the E step, we define the expectation

$$Q_L = E\Big[\log L(X, Z)\Big], \tag{6.20}$$

which starts with arbitrary values initially and repeatedly estimates their expectation. The aim in the M step is to use an optimizer to solve an optimization problem

$$\text{maximize } Q_L = E\Big[\log L(X, Z)\Big]. \tag{6.21}$$

The optimizer can be a gradient-based search algorithm.

The exact form of the likelihood function can be difficult to write, depending on the structure of the BBN. In some cases, the network structure may not be known in advance, which makes it extremely difficult. In this case, some heuristic and metaheuristic approaches may be needed to learn the network structures. Interested readers can consult a more advanced literature.

6.7 Data mining for big data

The big data science has become increasingly important nowadays, driven by the Internet, social media, and internet of things (IoT). Many applications are now dynamically data-driven. Comparing with traditional databases and data analytics, big data have some key characteristics, and thus the techniques required to cope with such big data are also more sophisticated [76].

6.7.1 Characteristics of big data

The main characteristics of big data can be summarized as 5 Vs, and they are: volume, velocity, variety, value, and veracity [66,76,78,99].

- Volume: the volume of data has increased dramatically in recent years, driven by the Internet, multimedia, and social media.
- Velocity: the rate of data accumulation is also increased dramatically. For example, it is estimated that there are about 20 trillion GB data added each year.
- Variety: the variety of data is also diverse, and data can be structured and unstructured from different types, sources, and media. For example, digital astronomy can have large datasets of images and sky survey images, but they are mainly structured data. In comparison, big data from social media, digital economy, and internet of things can collect a huge range of different data types.

In addition, the dimensionality of big data can be high, depending on many different factors, features or variables, either explicit or hidden.

The collection, storage and processing of such big data sets pose many challenges in terms of both hardware and software. However, whatever these challenges may be, we still have to try to extract some values from such big data [101].

- Value: The main aim of analysis and processing is to extract some useful features so as to gain insight into the data and to make predictions. Ideally, the ultimate aim is to understand the data so as to potentially predict future events.
- Veracity: Obviously, no matter how big the data may be, they can be incomplete with noise and uncertainty, subject to dynamic changes. Even so, the quality of the data and the subsequent analysis will have different accuracy and values.

All these characteristics, in addition to high dimensionality, make the data analysis a very challenging task, if not impossible.

6.7.2 Statistical nature of big data

Almost all data mining techniques use some statistical properties. For example, it is traditionally assumed that data are independent in the statistical sense. However, in reality, most data are not necessarily statistically independent, and thus statistical foundations for such methods may not be valid. Currently, there is no rigorous theory for methods that are based on statistically dependent data. Even so, researchers just assume whatever is most appropriate for the data so as to be able to use certain techniques to process and analyze the big data.

Thus, traditional methods are still used with some modifications to deal with big data, including regression, decision tree, hidden Markov models, neural networks, support vector machines, and others. However, from the statistical point of view, we should be aware of Bonferroni's principle when dealing with big data. If you try to calculate some expected number of occurrences of certain patterns and if this number is significantly larger than the number of real instances that you hope to find, then you can conclude that almost anything you find is false (which means it is an artifact in a statistical sense, rather than actual evidence). In order words, if you look too hard for interesting patterns that your data may support, you are bound to find false patterns. Thus care should be taken when interpreting data [96].

6.7.3 Mining big data

For big data, apart from other challenges such as storage and retrieval, one of the challenges is that the data to be processed is much larger than the main physical memory of the computer, and thus it is not possible to load all the data into a computer's main memory to process. Some kind of sampling and segment-by-segment processing may be needed before applying any algorithms.

Recent developments show that new methods may be more suitable for large datasets [96]. For example, the Bradley–Fayyad–Reina (BFR) algorithm [25] and Clustering Using REpresentative (CURE) algorithm have shown good results. For more detail about these two algorithms, we refer the readers to the books by Leskovec et al. [96], Bradley et al. [25], and Guha et al. [60].

In essence, the BFR algorithm is an extension of k-means to high-dimensional data. It assumes that data are distributed around centroids according to Gaussian distributions in Euclidean spaces. The essence of BFR algorithm can be outlined as follows.

1. Choose a small subset (by resampling) of the big data using either k-means or hierarchy methods to find the initial k clusters.
2. Take each chunk data (a subset) from the big dataset and do the following:
    - Assign the data points and summarize the clusters (see details further), then discard the data;
    - Compress and merge points that are close to one another (forming compressed sets);
    - Retain the data points that are not assigned or not close to one another (forming retained sets).

For any given N points (in a subset), calculate the SUM_i and SUMSQ_i, where SUM_i is the vector sum of data in the ith dimension, and SUMSQ_i is the sum of the squares of all the data points in the ith dimension.

Then, the centroid can be updated at SUM_i/N in the ith dimension, and its variance in the ith dimension can be estimated as $\text{SUMSQ}_i-(\text{SUM}_i/\text{N})^2$. The advantage of such dimension-by-dimension calculations is that both SUM_i and SUMSQ_i become simple sums when combining two clusters.

When deciding if a new point $\boldsymbol{x}_i$ is close enough to one of the k clusters, we can use the following rules: (1) add point $\boldsymbol{x}_i$ to a cluster if it has the centroid closest to $\boldsymbol{x}_i$, or (2) assign $\boldsymbol{x}_i$ to a cluster with the least Mahalanobis distance.

The Mahalanobis distance between $\boldsymbol{x} = (x_1, x_2, \ldots, x_n)$ and a cluster centroid $c = (c_1, c_2, \ldots, c_n)$ is

$$d_m = \sqrt{\sum_{i=1}^{n} \Big(\frac{x_i - c_i}{\sigma_i}\Big)^2}, \tag{6.22}$$

where σ_i is the standard deviation of the cluster in the ith dimension. Therefore this distance is a variance-based scaled distance.

For a subset of data points (a chunk from the big data), the detailed calculations are as follows:

- For data points that close to the centroid of a cluster, add these points to that cluster. Then, update the centroid and other metrics such as SUM_i and SUMSQ_i.
- For points that are not close to any centroid, cluster them, together with the retained sets. Then, merge any miniclusters when appropriate.
- For points that are assigned to a cluster (including any minicluster), update the centroid and other metrics (then discard such points).
- In the final stage (after going through all the subset of the data or loading the last chunk of the data), postprocess the retained sets and compressed sets by either assigning each point to the cluster of the nearest centroid or discarding them as outliers.

Though the BFR algorithm is efficient, it is mainly for data that are symmetric around clusters, and thus it cannot deal with S-shapes or rings effectively. For such complicated datasets, we can use another powerful algorithm, called the CURE (Clustering Using REpresentives) algorithm, which is a point-assignment algorithm in Euclidean spaces without any assumptions about the shape of clusters (unlike the normal distribution assumption in the BFR algorithm). As a result, this algorithm can deal with data clustering of odd shapes such as bends, S-shapes, and rings.

The main steps of the CURE algorithm can be summarized as follows [60]:

1. Choose a small sample (a small subset) of the data so as to be clustered in the main memory using any good methods such as k-means and hierarchy methods, which gives the initial clusters.
2. Select a small set of points from each cluster to be representative points (points should be as far from one another as possible).

3. Move each of the representative points a fixed fraction (typically 20%) of the distance between its location to the centroid of its cluster.
4. Merge two clusters if they have a pair of representation points (one from each cluster) that are sufficiently close.
5. Repeat the previous steps and merge until no more close cluster to merge.
6. Carry out point assignment for all remaining points in the big data.

Though this algorithm can be sufficiently efficient in practice, there is no guarantee for the global optimality. For large datasets, it is impractical to reach the true global optimality. Any suboptimal or sufficient good solutions can become acceptable in practical applications.

It is worth pointing out that the above clustering calculations have been based on the Euclidean distances. Obviously, non-Euclidean distances such as Jaccard similarity, Edit distance, and Hamming distance can be used, depending on the types of problems.

Recent trends tend to combine traditional algorithms with optimization algorithms that are based on swarm intelligence. The basic idea is to use optimization techniques to optimize the centroids and then use clustering methods such as k-means to carry out clustering. Recent studies suggest that such hybrid methods can produce very promising results [37]. For example, the firefly algorithm can be used to do clustering and classifications with superior performance.

For any methods to be efficient and useful, large matrices should be avoided, and there is no need to try every possible combination. The methods used to solve a large-scale problem should be efficient enough to produce good results in a practically acceptable time scale. However, in general, there is no guarantee that the global optimality can be found.

There are other methods such as GRGPF and BIRCH algorithms for clustering big datasets. In addition, algorithms and implementations can be parallelized to speed up these algorithms. The readers can consult a more advanced literature on these topics [96].

6.8 Notes on software

There are a vast array of software packages for data mining, and it is almost an impossible task to mention all such packages. Matlab, R, Python, Mathematica, and many others have implemented k-means, kNN, and various classifiers.

For example, R has `kmeans`, `kNN`, decision trees, and random forest packages. In addition, R can be paired with big databases such as Amazon Redshift and Google BigQuery. Python has `kmeans` and decision-tree classifiers. There are also Python libraries such as Pandas, NumPy, SciPy, and machine learning libraries such as Mlpy, Theano, Scikit-learn, and others. Matlab has all the functionalities such as `kmeans`, `fitcknn`, `fitctree`, and `fitrensemble`. More computer codes in Matlab can be found from Mathworks file exchanges, though not all files are well tested on the file exchange web site.

Furthermore, RapidMiner has all these classifiers implemented, including logistic regression, decision trees, and naive Bayesian classifier. It is worth mentioning that these algorithms, including random forests, have been implemented in other programming languages such as C++.

Support vector machine and regression

7

Contents

Support vector machines are a class of powerful tools, which become increasingly popular in classifications, data mining, pattern recognition, artificial intelligence, and optimization.

In many applications, the aim is separating some complex data into different categories. For example, in pattern recognition, we may need to simply separate different images into different classes, that is, to label them with categorical values. In other applications, we have to answer a yes–no question, which is a binary classification. If there are k different classes, then we can in principle first classify them into two classes, (say) class 1 and nonclass 1. We then focus on the nonclass 1 and divide it into two different classes, and so on.

Mathematically speaking, for a given set of scattered data, the objective is separating them into different regions/domains or categorical types. In the simplest case, the outputs are just class either A or B, that is, either $+1$ or -1.

7.1 Statistical learning theory

For the case of two-class classification, we have the learning examples or data as $(\boldsymbol{x}_i, y_i)$ where $i = 1, 2, \ldots, n$ and $y_i \in \{-1, +1\}$. The aim of such learning is to find a function $f_\beta(\boldsymbol{x})$ from allowable functions $\{f_\beta : \beta \in \Omega\}$ in the parameter space Ω such that

$$f_\beta(\boldsymbol{x}_i) \mapsto y_i \qquad (i = 1, 2, \ldots, n) \tag{7.1}$$

and such that the expected risk $E(\beta)$ is minimal. The latter is defined as

$$E(\beta) = \frac{1}{2} \int |f_\beta(x) - y| dP(\boldsymbol{x}, y), \tag{7.2}$$

Introduction to Algorithms for Data Mining and Machine Learning. https://doi.org/10.1016/B978-0-12-817216-2.00014-4

where $P(\boldsymbol{x}, y)$ is an unknown probability distribution, which makes it impossible to calculate $E(\beta)$ directly. A simple approach is using the so-called empirical risk

$$E_p(\beta) \approx \frac{1}{n} \sum_{i=1}^{n} \frac{1}{2} \big| f_\beta(\boldsymbol{x}_i) - y_i \big|. \tag{7.3}$$

A main drawback of this approach is that a small risk or error on the training set does not necessarily guarantee a small error on prediction if the number n of training data is small.

In the framework of structural risk minimization and statistical learning theory, there exists an upper bound for such errors. For a given probability of at least $1 - p$, the Vapnik bound for the errors can be written as

$$E(\beta) \le R_p(\beta) + \phi\Big(\frac{h}{n}, \frac{\log(p)}{n}\Big), \tag{7.4}$$

where

$$\phi\Big(\frac{h}{n}, \frac{\log(p)}{n}\Big) = \sqrt{\frac{1}{n}\Big[h(\log\frac{2n}{h} + 1) - \log(\frac{p}{4})\Big]}. \tag{7.5}$$

Here h is a parameter, often referred to as the Vapnik–Chervonenskis dimension (or simply VC-dimension) [143]. This dimension describes the capacity for prediction of the function set f_β. In the simplest binary classification with only two values of $+1$ and -1, h is essentially the maximum number of points that can be classified into two distinct classes in all possible 2^h combinations.

7.2 Linear support vector machine

The basic idea of classification is trying to separate different samples into different classes. For binary classification such as the diamonds and circles as shown in Fig. 7.1, we intend to construct a hyperplane

$$\boldsymbol{w} \cdot \boldsymbol{x} + b = 0 \tag{7.6}$$

so that these samples can be divided into two classes with all the triangles on one side and the spheres on the other side. Here the normal vector $\boldsymbol{w}$ should have the same size as $\boldsymbol{x}$, and they can be determined using the data, though the method of determining them is not straightforward. This requires the existence of a hyperplane; otherwise, this approach will not work. In this case, we have to use other methods.

It is worth pointing out that $\boldsymbol{w} \cdot \boldsymbol{x} + b = 0$ can also be written as

$$\boldsymbol{w}^T \boldsymbol{x} + b = 0, \tag{7.7}$$

but the dot product form of $\boldsymbol{w} \cdot \boldsymbol{x}$ explicitly highlights the nature of a hyperplane governed by the normal direction $\boldsymbol{w}$.

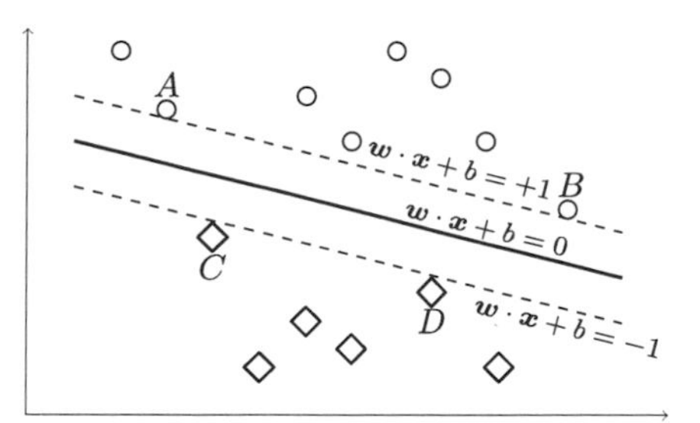

Figure 7.1 Hyperplane, maximum margins, and a linear support vector machine (SVM).

If we can construct such a hyperplane, we should construct two hyperplanes (shown as dashed lines in Fig. 7.1) so that the two hyperplanes should be as far away as possible and no samples should be between these two planes. Mathematically, this is equivalent to two equations

$$\boldsymbol{w} \cdot \boldsymbol{x} + b = +1 \tag{7.8}$$

and

$$\boldsymbol{w} \cdot \boldsymbol{x} + b = -1. \tag{7.9}$$

It is worth pointing out that these equations in the literature are also written as

$$y(\boldsymbol{x}) = f(\boldsymbol{x}) = < \boldsymbol{w}, \boldsymbol{x} > + b = \pm 1. \tag{7.10}$$

From these two equations it is straightforward to verify that the normal (perpendicular) distance d between these two hyperplanes is related to the norm $||\boldsymbol{w}||$ via

$$d = \frac{2}{||\boldsymbol{w}||}. \tag{7.11}$$

The main objective of constructing these two hyperplanes is maximizing the distance or the margin between the two planes. The maximization of d is equivalent to the minimization of $||w||$ or, more conveniently, $||w||^2/2$. Here $||w||$ is the standard L_2-norm defined earlier in Chapter 2.

From the optimization point of view, the maximization of margins can be written as

$$\text{minimize } \frac{1}{2}||\boldsymbol{w}||^2 = \frac{1}{2}||\boldsymbol{w}||_2^2 = \frac{1}{2}(\boldsymbol{w} \cdot \boldsymbol{w}). \tag{7.12}$$

If we can classify all the samples completely, then for any sample $(\boldsymbol{x}_i, y_i)$, $i = 1, 2, \ldots, n$, we have

$$\boldsymbol{w} \cdot \boldsymbol{x}_i + b \geq +1 \qquad \text{if } (\boldsymbol{x}_i, y_i) \in \text{one class}, \tag{7.13}$$

and

$$\boldsymbol{w} \cdot \boldsymbol{x}_i + b \leq -1 \qquad \text{if } (\boldsymbol{x}_i, y_i) \in \text{the other class.} \tag{7.14}$$

As $y_i \in \{+1, -1\}$, these two equations can be combined as

$$y_i(\boldsymbol{w} \cdot \boldsymbol{x}_i + b) \geq 1 \qquad (i = 1, 2, \ldots, n). \tag{7.15}$$

However, in reality, it is not always possible to construct such a separating hyperplane. A very useful approach is using nonnegative slack variables

$$\eta_i \geq 0 \qquad (i = 1, 2, \ldots, n), \tag{7.16}$$

so that

$$y_i(\boldsymbol{w} \cdot \boldsymbol{x}_i + b) \geq 1 - \eta_i \qquad (i = 1, 2, \ldots, n). \tag{7.17}$$

Now the optimization problem for the support vector machine becomes

$$\text{minimize } \Psi = \frac{1}{2}||\boldsymbol{w}||^2 + \lambda \sum_{i=1}^{n} \eta_i, \tag{7.18}$$

subject to

$$y_i(\boldsymbol{w} \cdot \boldsymbol{x}_i + b) \geq 1 - \eta_i, \tag{7.19}$$

$$\eta_i \geq 0 \qquad (i = 1, 2, \ldots, n), \tag{7.20}$$

where $\lambda > 0$ is a penalty parameter to be chosen appropriately. Here, the term $\sum_{i=1}^{n} \eta_i$ is essentially a measure of the upper bound of the number of misclassifications on the training data.

By using Lagrange multipliers $\alpha_i \geq 0$ we can rewrite the constrained optimization into an unconstrained version, and we have

$$L = \frac{1}{2}||\boldsymbol{w}||^2 + \lambda \sum_{i=1}^{n} \eta_i - \sum_{i=1}^{n} \alpha_i [y_i(\boldsymbol{w} \cdot \boldsymbol{x}_i + b) - (1 - \eta_i)]. \tag{7.21}$$

From this we can write the Karush–Kuhn–Tucker conditions as

$$\frac{\partial L}{\partial \boldsymbol{w}} = \boldsymbol{w} - \sum_{i=1}^{n} \alpha_i y_i \boldsymbol{x}_i = 0, \tag{7.22}$$

$$\frac{\partial L}{\partial b} = -\sum_{i=1}^{n} \alpha_i y_i = 0, \qquad y_i(\boldsymbol{w} \cdot \boldsymbol{x}_i + b) - (1 - \eta_i) \geq 0, \tag{7.23}$$

$$\alpha_i [y_i(\boldsymbol{w} \cdot \boldsymbol{x}_i + b) - (1 - \eta_i)] = 0 \qquad (i = 1, 2, \ldots, n), \tag{7.24}$$

$$\alpha_i \geq 0, \qquad \eta_i \geq 0 \qquad (i = 1, 2, \ldots, n). \tag{7.25}$$

From the first KKT condition, we get

$$\boldsymbol{w} = \sum_{i=1}^{n} y_i \alpha_i \boldsymbol{x}_i. \tag{7.26}$$

It is worth pointing out here that only the nonzero coefficients α_i contribute to the overall solution. This comes from the KKT condition (7.24), which implies that when $\alpha_i \neq 0$, inequality (7.19) must be satisfied exactly, whereas $\alpha_i = 0$ means that the inequality is automatically met. Therefore, only the corresponding training data $(\boldsymbol{x}_i, y_i)$ with $\alpha_i > 0$ can contribute to the solution, and thus such $\boldsymbol{x}_i$ form the support vectors (hence, the name support vector machine). For example, the position vectors for the four points (A, B, C, D) in Fig. 7.1 are support vectors. All the other data with $\alpha_i = 0$ become irrelevant [34,143].

There is a dual problem for this SVM optimization problem [33,133,143], and it can be shown that the solution for α_i can be found by solving the following quadratic programming:

$$\text{maximize} \sum_{i=1}^{n} \alpha_i - \frac{1}{2} \sum_{i,j=1}^{n} \alpha_i \alpha_j y_i y_j (\boldsymbol{x}_i \cdot \boldsymbol{x}_j), \tag{7.27}$$

subject to

$$\sum_{i=1}^{n} \alpha_i y_i = 0, \tag{7.28}$$

$$0 \leq \alpha_i \leq \lambda \qquad (i = 1, 2, \ldots, n). \tag{7.29}$$

From the coefficients α_i we can write the final classification or decision function as

$$f(\boldsymbol{x}) = \text{sign}\Big[\sum_{i=1}^{n} \alpha_i y_i (\boldsymbol{x} \cdot \boldsymbol{x}_i) + b\Big], \tag{7.30}$$

where $\text{sign}(x)$ is the classic sign function: $\text{sign}(x) = +1$ if $x > 0$, -1 if $x < 0$, and 0 if $x = 0$.

7.3 Kernel functions and nonlinear SVM

In reality, most problems are nonlinear, and the previous linear SVM cannot be used. Ideally, we should find some nonlinear transformation ϕ such that the data can be mapped onto a high-dimensional feature space where classification becomes linear (see Fig. 7.2). The transformation should be chosen in a certain way so that their dot product leads to a kernel-style function

$$K(\boldsymbol{x}, \boldsymbol{x}_i) = \phi(\boldsymbol{x}) \cdot \phi(\boldsymbol{x}_i), \tag{7.31}$$

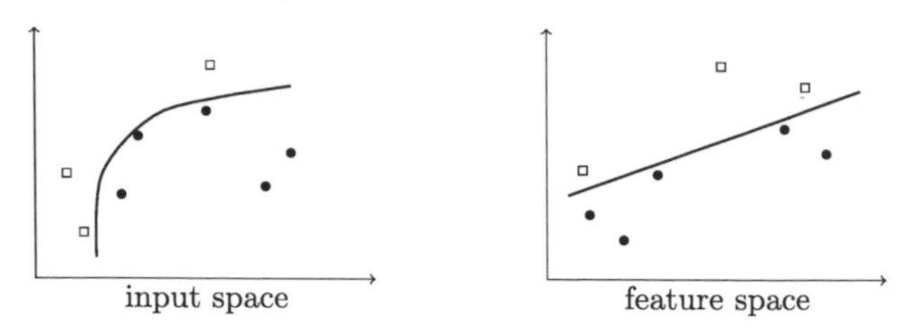

Figure 7.2 Kernels and nonlinear transformation.

which enables us to write our decision function as

$$f(\boldsymbol{x}) = \text{sign}\Big[\sum_{i=1}^{n} \alpha_i y_i K(\boldsymbol{x}, \boldsymbol{x}_i) + b\Big]. \tag{7.32}$$

From the theory of eigenfunctions we know that it is possible to expand functions in terms of eigenfunctions [43]. In fact, we do not need to know such transformations; we can directly use kernel functions $K(\boldsymbol{x}, \boldsymbol{x}_i)$ to complete this task. This is the so-called kernel function trick. Now the main task is choosing a suitable kernel function for a given problem.

For most problems concerning a nonlinear support vector machine, we can use

$$K(\boldsymbol{x}, \boldsymbol{x}_i) = (\boldsymbol{x} \cdot x_i)^d \tag{7.33}$$

for polynomial classifiers and

$$K(\boldsymbol{x}, \boldsymbol{x}_i) = \tanh[k(\boldsymbol{x} \cdot \boldsymbol{x}_i) + \Theta)] \tag{7.34}$$

for neural networks. The most widely used kernel is the Gaussian radial basis function (RBF)

$$\begin{aligned} K(\boldsymbol{x}, \boldsymbol{x}_i) &= \exp\Big[- ||\boldsymbol{x} - \boldsymbol{x}_i||^2/(2\sigma^2)\Big] \\ &= \exp\Big[- \gamma ||\boldsymbol{x} - \boldsymbol{x}_i||^2\Big] = \exp[-\gamma r^2] \end{aligned} \tag{7.35}$$

for nonlinear classifiers. Here $r = ||\boldsymbol{x} - \boldsymbol{x}_i||$. This kernel can easily be extended to any high dimensions. Here σ^2 is the variance, and $\gamma = 1/2\sigma^2$ is a constant. In fact, γ is a hyperparameter, which needs to be tuned for each support vector machine.

Following a similar procedure as discussed earlier for linear SVMs [86,133], we can obtain the coefficients α_i by solving the following optimization problem:

$$\text{maximize} \sum_{i=1}^{n} \alpha_i - \frac{1}{2}\alpha_i \alpha_j y_i y_j K(\boldsymbol{x}_i, \boldsymbol{x}_j). \tag{7.36}$$

It is worth pointing out that when the matrix $\boldsymbol{A} = y_i y_j K(\boldsymbol{x}_i, \boldsymbol{x}_j)$ is a symmetric positive definite matrix, the above maximization problem becomes a quadratic programming problem [22] and can thus be solved efficiently by quadratic programming techniques.

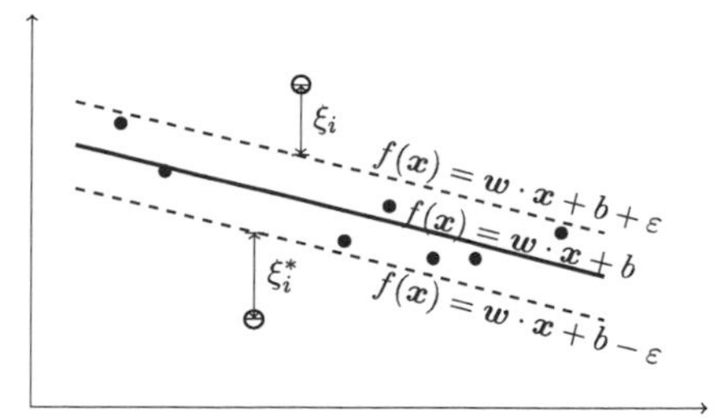

Figure 7.3 Support vector regression (SVR).

There are many easily available software packages (commercial or open source), so we will not provide any discussion on the implementation. In addition, some methods and their variants are still an area of active research. Interested readers can consult a more advanced literature.

7.4 Support vector regression

Support vector machines for classification work well for discrete labels. In case of a continuous dependent variable, SVMs will not work well because the problem is a regression problem. However, the basic ideas of SVMs can be modified to do regression. In this case, we have the support vector regression (SVR) method [8,40,133].

For a given tolerance ε, the ε-insensitive support vector regression (SVR) is finding a function $y = f(\boldsymbol{x}) = \boldsymbol{w}^T\boldsymbol{x} + b = \boldsymbol{w} \cdot \boldsymbol{x} + b$ such that all the data points are within a strip bounded by two hyperplanes (see Fig. 7.3); that is, all the data points $(\boldsymbol{x}_i, y_i)$ $(i = 1, 2, \ldots, n)$ should deviate at most ε (as a vertical distance) from the regression targets.

Mathematically, this regression problem can be written as

$$\text{minimize } \frac{1}{2}||\boldsymbol{w}||^2, \tag{7.37}$$

subject to

$$|y_i - \boldsymbol{w} \cdot \boldsymbol{x}_i - b| \leq \varepsilon \tag{7.38}$$

for all the n data point pairs $(\boldsymbol{x}_i, y_i)$, $i = 1, 2, \ldots, n$. This is equivalent to

$$\text{minimize } \frac{1}{2}||\boldsymbol{w}||^2, \tag{7.39}$$

subject to

$$\begin{cases} y_i - \boldsymbol{w} \cdot \boldsymbol{x}_i - b \leq \varepsilon, \\ \boldsymbol{w} \cdot \boldsymbol{x}_i + b - y_i \leq \varepsilon. \end{cases} \tag{7.40}$$

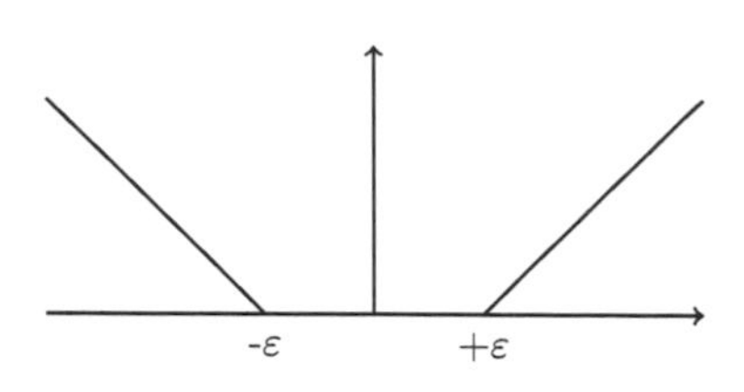

Figure 7.4 The ε-insensitive loss function.

This is an optimization problem, and in principle it can be solved using any proper optimization technique. However, depending on the choice of ε, there may not be any solution such that all the data points lie within $\pm\varepsilon$ from the regression line. A relaxation extension is to allow some errors by using some slack variables $\xi_i \geq 0$ (for the upper boundary) and $\xi_i^* \geq 0$ (for the lower boundary); see Fig. 7.3. All the values of these slack variables must be nonnegative ($\xi_i, \xi_i^* \geq 0$). The idea is to minimize $||\boldsymbol{w}||^2$ and also such errors. Thus, the regression optimization problem can be written as

$$\text{minimize } \frac{1}{2}||\boldsymbol{w}||^2 + \lambda \sum_{i}^{n} (\xi_i + \xi_i^*), \tag{7.41}$$

subject to

$$\begin{cases} y_i - \boldsymbol{w} \cdot \boldsymbol{x}_i - b \leq \varepsilon + \xi_i, \\ \boldsymbol{w} \cdot \boldsymbol{x}_i + b - y_i \leq \varepsilon + \xi_i^*, \\ \xi_i \geq 0, \quad \xi_i^* \geq 0 \ (i = 1, 2, \ldots, n). \end{cases} \tag{7.42}$$

Here the penalty parameter $\lambda > 0$ controls the tradeoff between the flatness of the regression model $f(\boldsymbol{x})$ and the deviation errors [40].

In the context of machine learning, this is equivalent to the so-called ε-insensitive loss function

$$|\xi|_\varepsilon = \begin{cases} 0 & \text{if } |\xi| < \varepsilon, \\ |\xi| - \varepsilon & \text{otherwise.} \end{cases} \tag{7.43}$$

Here we have used the same notation as that in the tutorial by Smola and Schölkopf [133]. This is shown in Fig. 7.4.

This regression problem can be solved by using Lagrangian multipliers $\eta_i, \eta_i^*, \alpha_i, \alpha_i^*$ to incorporate the inequality constraints. Thus, we have the Lagrangian

$$\begin{aligned} L = \frac{1}{2}||\boldsymbol{w}||^2 + \lambda \sum_{i=1}^{n} (\xi_i + \xi_i^*) - \sum_{i=1}^{n} (\eta_i \xi_i + \eta_i^* \xi_i^*) \\ - \sum_{i=1}^{n} \alpha_i (\varepsilon + \xi_i - y_i + \boldsymbol{w} \cdot \boldsymbol{x}_i + b) - \sum_{i=1}^{n} \alpha_i^* (\varepsilon + \xi_i^* + y_i - \boldsymbol{w} \cdot \boldsymbol{x}_i - b). \end{aligned} \tag{7.44}$$

As the regression model is linear, this is essentially equivalent to an unconstrained quadratic programming problem.

The optimality can be achieved by

$$\frac{\partial L}{\partial \boldsymbol{w}} = 0, \quad \frac{\partial L}{\partial b} = 0, \quad \frac{\partial L}{\partial \xi} = 0, \quad \frac{\partial L}{\partial \xi_i^*} = 0, \tag{7.45}$$

and the duality feasibility conditions

$$\eta_i \geq 0, \quad \eta_i^* \geq 0 \quad (i = 1, 2, \ldots, n). \tag{7.46}$$

We have

$$\frac{\partial L}{\partial \boldsymbol{w}} = \boldsymbol{w} - \sum_{i=1}^{n} \alpha_i \boldsymbol{x}_i + \sum_{i=1}^{n} \alpha_i^* \boldsymbol{x}_i = 0, \tag{7.47}$$

which gives

$$\boldsymbol{w} = \sum_{i=1}^{n} (\alpha_i - \alpha_i^*) \boldsymbol{x}_i. \tag{7.48}$$

In addition, we have

$$\frac{\partial L}{\partial b} = -\sum_{i=1}^{n} \alpha_i + \sum_{i=1}^{n} \alpha_i^* = \sum_{i=1}^{n} (\alpha_i^* - \alpha_i) = 0. \tag{7.49}$$

Furthermore, we have

$$\frac{\partial L}{\partial \xi_i} = \lambda - \eta_i - \alpha_i = 0, \quad \frac{\partial L}{\partial \xi_i^*} = \lambda - \eta_i^* - \alpha_i^* = 0, \tag{7.50}$$

leading to

$$\eta_i = \lambda - \alpha_i, \quad \eta_i^* = \lambda - \alpha_i^*. \tag{7.51}$$

Since $\eta_i, \eta_i^* \geq 0$, these two conditions mean that

$$\alpha_i \in [0, \lambda], \quad \alpha_i^* \in [0, \lambda]. \tag{7.52}$$

Similar to the SVM, kernel tricks can also be used to deal with nonlinear support vector regression. We refer the interested readers to the more advanced literature such as Vapnik's 1995 book [143] and the tutorial by Smola and Schölkopf [133].

7.5 Notes on software

Both SVM and SVR are widely used, so many software packages have implemented both methods. Matlab, R, Python, Mathematica, and Maple all have such functionalities. For example, Matlab has `fitcsvm` and `fitrsvm` for classification and regression, respectively, R has `svm`, whereas the Scikit-Learn for Python has `svm` with all algorithm variants. In addition, SVM and SVR are also been implemented in C++, Java, and other machine learning packages.

Neural networks and deep learning

Contents

Machine learning algorithms are a class of sophisticated optimization algorithms, including both supervised learning and unsupervised learning algorithms. In general, there are a diverse range of algorithms in this category, and they are classification and clustering algorithms, regression, decision trees, artificial neural networks, support vector machines, Bayesian networks, Boltzmann machines, natural language processing, deep belief networks, and others [144,105]. We have already introduced some of these techniques. In this chapter, we mainly focus on artificial neural networks (ANN) and deep learning (DL).

Many applications use ANNs, which is especially true in artificial intelligence. The fundamental idea of ANNs is learning from data and making predictions using a network of connected neurons arranged into a layered structure. For a given set of input data, a neural network maps the input data into some outputs. The relationships between the inputs and outputs are quite complicated, and it is usually impossible to express such relationships in any exact or analytical form. By comparing the outputs with the true outputs, the system can adjust its weights so as to better match its outputs. If there is a sufficient number of data, then the network can become well trained, and thus the trained network can make predictions for new data.

8.1 Learning

In general, learning in the broad context of machine learning can be divided into three categories: supervised learning, unsupervised learning, and reinforcement learning.

Introduction to Algorithms for Data Mining and Machine Learning. https://doi.org/10.1016/B978-0-12-817216-2.00015-6

Supervised learning use data with known labels or classes. Regression is an example of supervised learning: the target outputs (y) are real numbers via a model or model class $y = f(\boldsymbol{x}, \boldsymbol{\alpha})$, where $\boldsymbol{\alpha}$ is a model parameter vector. In addition, classification is also supervised learning. For example, classification using the support vector machine belongs to this category [4,17].

Clustering is a good example of unsupervised learning. There is no need to have labeled data, and learning is figuring out the internal structures and representation of the data.

Reinforcement learning is based on some scalar reward objective to be maximized, based on the data and some future rewards. However, this type of learning is limited to deal with cases where the number of key parameters is not huge (e.g., a few dozens).

It is worth pointing out that the learning in neural networks can be supervised or unsupervised, depending on the type of data. Using labeled data with target outputs to train neural networks is supervised learning. Otherwise, if there are no target outputs, then the learning becomes unsupervised.

In addition to the traditional categories of learning, other forms and new types of learning exist in the literature. For example, semisupervised learning is the learning of largely using a lot of unlabeled data with a few labeled data, where both labeled and unlabeled data are assumed to be drawn from the same probability distributions. In essence, semisupervised learning implicitly uses the assumption that unlabeled data can be labeled with the same labels used by labeled data for classification tasks.

Now let us start with artificial neural networks. Then we will move onto more sophisticated topics about Boltzmann machines and deep learning.

8.2 Artificial neural networks

A neural network consists of many connected neurons, and the behavior of a neuron is defined by a neuron model.

8.2.1 Neuron models

The basic mathematical model of an artificial neuron was first proposed by W. McCulloch and W. Pitts in 1943, and this fundamental model is referred to as the McCulloch–Pitts model. Other models and neural networks are based on it [64,65,124].

An artificial neuron with n inputs or impulses and an output y is activated if the signal strength reaches a certain threshold θ. Each input has a corresponding weight w_i (see Fig. 8.1). The output of this neuron is given by

$$y = f(x), \quad x = \sum_{i=1}^{n} w_i u_i, \tag{8.1}$$

where the weighted sum x is the total signal strength, and $f(x) = f(\sum_i u_i)$ is the so-called activation function, which usually depends on a threshold parameter θ or

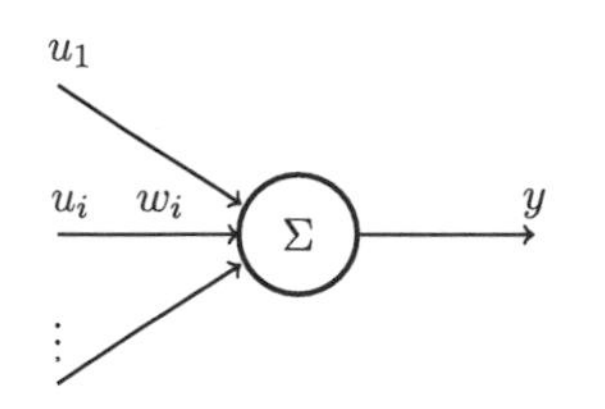

Figure 8.1 A simple neuron model.

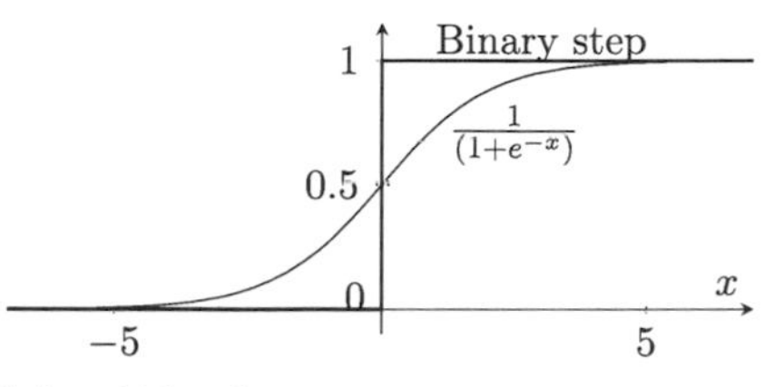

Figure 8.2 Binary step and sigmoid function.

some control parameter. The simplest form of this function can be $f(x) = x$, called the identity activation function, which simply passes the input as the output.

A binary step function takes the following form:

$$f(x) = \begin{cases} 1 & \text{if } x \geq \theta, \\ 0 & \text{if } x < \theta. \end{cases} \tag{8.2}$$

We can see that the output is only activated to a nonzero value such as unity if the overall signal strength is greater than the threshold θ.

The step activation function is binary and has discontinuity; sometimes, for activation, it is easier to use the nonlinear smooth function, called the sigmoid function,

$$S(x) = \frac{1}{1 + e^{-x}}, \tag{8.3}$$

which approaches 1 as $x \to +\infty$ and becomes 0 as $x \to -\infty$ (see Fig. 8.2).

The sigmoid function is also called the logistic activation function or soft step function in the literature. An interesting property of this function is

$$S'(x) = S(x)[1 - S(x)]. \tag{8.4}$$

8.2.2 Activation models

There are many different models for activation, and they are often called activation functions. Different models approximate and smoothen activation behavior slightly differently, including the rectified linear model, hyperbolic tangent model, the exponential linear model, and others. Here we introduce a few commonly used activation functions.

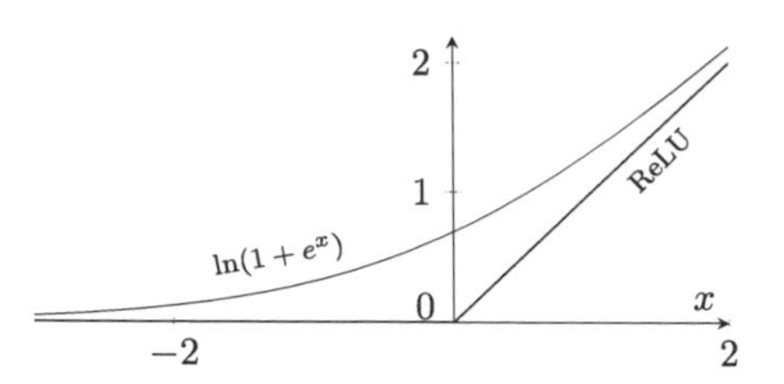

Figure 8.3 ReLU = $\max(0, x)$ and SoftPlus = $\ln(1 + e^x)$ functions.

The rectified linear model, also called the rectified linear unit (ReLU), can be written as

$$f(x) = x^{+1} = \max\{0, x\} = \begin{cases} x & \text{if } x > 0, \\ 0 & \text{otherwise.} \end{cases} \tag{8.5}$$

The ReLU is quite widely used in deep learning due to its simplicity. In addition, it has a constant gradient 1 for any value of $x > 0$, which is a desirable property. In contrast, the gradient of a sigmoid function is approaching zero when x is very large.

The ReLU function can be approximated by a smoother function, called SoftPlus,

$$f(x) = \log(1 + e^x), \tag{8.6}$$

whose derivative is $f'(x) = e^x/(1 + e^x) = S(x)$. Both ReLU and SoftPlus functions are shown in Fig. 8.3.

Though the ReLU is very widely used, it may block some units due to its zero gradient when $x < 0$. A natural extension of this is using a small gradient for $x < 0$, allowing propagation to pass through (leaking). In this case, ReLU is extended to the parametric rectified linear unit (PReLU), more often called the leaky rectified linear unit (Leaky ReLU), with parameter α (typically, $\alpha = 0.01$):

$$f(x) = \begin{cases} \alpha x & \text{if } x < 0, \\ x & \text{if } x \geq 0. \end{cases} \tag{8.7}$$

In addition, the exponential linear unit (ELU) is defined by

$$f(x) = \begin{cases} \alpha(e^x - 1) & \text{if } x < 0, \\ x & \text{if } x \geq 0. \end{cases} \tag{8.8}$$

The hyperbolic tangent activation can be written as

$$f(x) = \tanh(x) = \frac{e^x - e^{-x}}{e^x + e^{-x}} = \frac{1 - e^{-2x}}{1 + e^{-2x}} = \frac{2}{1 + e^{-2x}} - 1, \tag{8.9}$$

which approaches to $+1$ as $x \to \infty$ and -1 as $x \to -\infty$. Another related activation function is the arctan function $f(x) = \tan^{-1}(x)$.

In case of multiple m inputs from a previous layer, the so-called softmax activation is often used to convert to probabilities to m classes. The probability for class i is given by

$$P_i = \frac{e^{x_i}}{\sum_{j=1}^{m} e^{x_j}}, \quad i = 1, 2, \ldots, m, \tag{8.10}$$

which is essentially the same as the softmax regression. Interestingly, its Jacobian $\boldsymbol{J} = [J_{ij}]$ can be written as

$$J_{ij} = \frac{\partial P_i}{\partial x_j} = P_i(\delta_{ij} - P_j) \quad (i, j = 1, 2, \ldots, m), \tag{8.11}$$

where δ_{ij} is the Kronecker delta function, that is, $\delta_{ij} = 1$ if $i = j$ and $\delta_{ij} = 0$ if $i \neq j$.

Researchers have also designed other activation functions such as adaptive piecewise functions. One question is which activation function should be used, and the answer depends on many factors such as type of problems, the structure of the networks, and other factors. In general, ReLU, Leaky ReLU, and Softmax are among the most widely used. We refer the interested readers to the more advanced literature [10,105].

8.2.3 Artificial neural networks

A single neuron can only perform a simple task, on or off. Complex functions can be designed and performed using a network of interconnecting neurons or perceptrons. The structure of a network can be complicated, and one of the most widely used is arranging them in a layered structure, with an input layer, an output layer, and one or more hidden layers (see Fig. 8.4). The connection strength between two neurons is represented by its corresponding weight. It is worth pointing out that there can be multiple hidden layers for an artificial neural network; however, for simplicity of discussions, we only draw one hidden layer in Fig. 8.4. The method for updating the weights is the same for multiple hidden layers, especially for the case of backward propagation algorithm to be introduced later.

Some artificial neural networks (ANNs) can perform complex tasks and simulate complex mathematical models, even if there is no explicit functional form mathematically. Neural networks have developed over last few decades and have been applied in almost all areas of science and engineering.

The construction of a neural network involves the estimation of the suitable weights of a network system with some training/known data sets. The task of training is finding a set of suitable weights w_{ij} so that the neural networks not only can best-fit the known data, but also can predict outputs for new inputs. A good artificial neural network should be able to minimize both errors simultaneously, the fitting/learning errors and the prediction errors.

The errors can be defined as the differences between the calculated (or predicted) output v_k and real output y_k for all N output neurons in the least-square sense:

$$E = \frac{1}{2} \sum_{k=1}^{N} (v_k - y_k)^2. \tag{8.12}$$

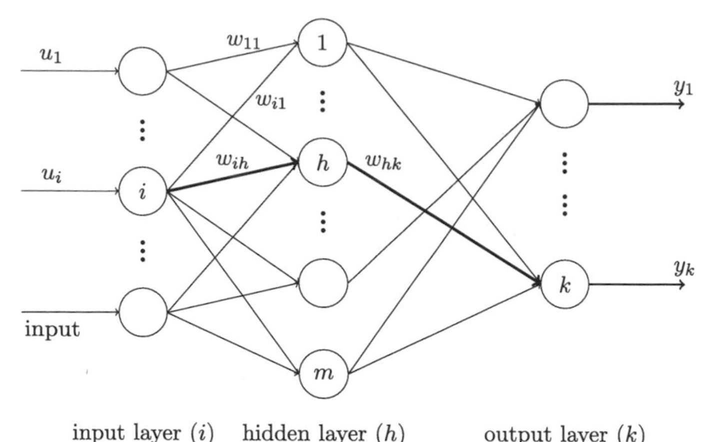

Figure 8.4 Schematic representation of a simple feedforward neural network with n_i inputs, m hidden nodes, and N outputs.

Here the output v_k is a function of inputs, activation, and weights. This can be also be written as

$$E = \frac{1}{2}||\boldsymbol{v} - \boldsymbol{y}||_2^2, \tag{8.13}$$

where $\boldsymbol{v}$ is the output vector, and $\boldsymbol{y}$ is the real or desired output vector. Here the error function E is often called the loss function of the artificial neural network. The main aim of training a neural network is minimizing the loss or training error and also maximizing the prediction accuracy.

To minimize this error, we can in principle use appropriate optimization techniques to find the solutions of the weights. A simple and yet efficient technique is the steepest descent method. For any initial random weights, the weight increment for w_{hk} is

$$\Delta w_{hk} = -\eta \frac{\partial E}{\partial w_{hk}} = -\eta \frac{\partial E}{\partial v_k} \frac{\partial v_k}{\partial w_{hk}}, \tag{8.14}$$

which follows the basic chain rule of differentiation. Here $0 < \eta \le 1$ is the learning rate. In a particular case, we can use $\eta = 1$ for discussions.

By starting with the neurons or nodes on the output layer in Fig. 8.4, we have

$$S_k = \sum_{h=1}^{m} w_{hk} v_h \qquad (k = 1, 2, \ldots, N), \tag{8.15}$$

and

$$v_k = f(S_k) = \frac{1}{1 + e^{-S_k}}. \tag{8.16}$$

From $f' = f(1 - f)$, we can obtain

$$\frac{\partial v_k}{\partial w_{hk}} = \frac{\partial v_k}{\partial S_k} \frac{\partial S_k}{\partial w_{hk}} = v_k(1 - v_k)v_h \tag{8.17}$$

and

$$\frac{\partial E}{\partial v_k} = (v_k - y_k). \tag{8.18}$$

Therefore we have

$$\Delta w_{hk} = -\eta \delta_k v_h, \tag{8.19}$$

where

$$\delta_k = v_k(1 - v_k)(v_k - y_k). \tag{8.20}$$

In general, based on (8.14), the increment for any weight w_{ij} related to errors is given by

$$\Delta w_{ij} = -\eta \frac{\partial E}{\partial w_{ij}}, \tag{8.21}$$

which can be written compactly as the iterative formula

$$\boldsymbol{w}^{t+1} = \boldsymbol{w}^t - \eta \nabla E(\boldsymbol{w}^t), \tag{8.22}$$

where ∇E is the gradient vector. However, for large-scale problems with many outputs, the computation of the gradient vectors can be very expensive, and thus some iterative or propagation formulas are preferred.

It is worth pointing out that the topology of a neural network is also important as the ways of arranging neurons will influence the algorithm used. Different topologies have different connections and thus different weights. Here we have used a common feedforward structure with neurons on the previous layer (on the left) affecting the neurons on the next layer (on the right), but the neurons on the right cannot affect the neurons on the left. Thus the inputs are fed forward to the outputs, and this structure allows an efficient implementation of the back propagation to be introduced in the next section.

For a given structure of the neural network, there are many ways of calculating weights by supervised learning. One of the simplest and widely used methods is using the back propagation algorithm for training neural networks, often called back propagation neural networks (BPNNs).

8.3 Back propagation algorithm

The basic idea of a BPNN is to start from the output layer and propagate backward so as to estimate and update the weights. This process is carried out in an iterative manner, until a predefined stopping criterion is met.

From any initial random weighting matrices w_{ih} (for connecting the input nodes to the hidden layer) and w_{hk} (for connecting the hidden layer to the output nodes) we can calculate the outputs of the hidden layer v_h:

$$v_h = \frac{1}{1+\exp[-\sum_{i=1}^{n_i} w_{ih} u_i]} \qquad (h = 1, 2, \ldots, m), \tag{8.23}$$

and the outputs for the output nodes

$$v_k = \frac{1}{1+\exp[-\sum_{h=1}^{m} w_{hk} v_h]} \qquad (k = 1, 2, \ldots, N). \tag{8.24}$$

The errors for the output nodes are given by

$$\delta_k = v_k(1 - v_k)(v_k - y_k) \qquad (k = 1, 2, \ldots, N), \tag{8.25}$$

where y_k $(k = 1, 2, \ldots, N)$ are the data (real outputs), for the given values of inputs u_i $(i = 1, 2, \ldots, n_i)$. Similarly, the errors for the hidden nodes can be written as

$$\delta_h = v_h(1 - v_h) \sum_{k=1}^{N} w_{hk} \delta_k \qquad (h = 1, 2, \ldots, m). \tag{8.26}$$

If we use a similar gradient descent algorithm, the updating formulas for weights at iteration t are

$$w_{hk}^{t+1} = w_{hk}^{t} - \eta \delta_k v_h \tag{8.27}$$

and

$$w_{ih}^{t+1} = w_{ih}^{t} - \eta \delta_h u_i, \tag{8.28}$$

where $0 < \eta \leq 1$ is the learning rate. Here we can see that the weight increments are

$$\Delta w_{ih} = -\eta \delta_h u_i, \tag{8.29}$$

which has a similar form to the earlier formula for w_{hk}.

Again, we can write all this in the following general formula:

$$\boldsymbol{w}^{t+1} = \boldsymbol{w}^{t} - \eta \nabla E(\boldsymbol{w}^{t}), \quad E = \frac{1}{2} \sum_{k=1}^{N} \Big[v_k(\boldsymbol{w}) - y_k \Big]^2, \tag{8.30}$$

which applies iteratively to each layer, propagating backward for multilayered networks.

It is worth pointing out that the error model we used is an L_2-norm (i.e., $||\boldsymbol{v} - \boldsymbol{y}||_2^2$) in terms of differences of the outputs and targets. There are other error models, often called loss functions for neural networks.

In a very particular case where the target outputs are the same as inputs (that is, using a neural network to learn and fitting the inputs to themselves), this becomes an autoencoder network. In this case, Hinton and Salakhutdinov [70] in 2006 used a multilayer neural network with a small central layer to reconstruct high-dimensional input vectors. The weights were adjusted and fine-tuned by gradient descent with a pretraining procedure to obtain better initial weights. They showed that their deep autoencoder network can learn low-dimensional representations that work much better than the principal component analysis, which can effectively reduce the dimensionality of data.

8.4 Loss functions in ANN

The loss function we have discussed so far is mainly the residual errors in terms of an L_2-norm as given in Eq. (8.13). There are many other forms of the loss functions in the literature, which may be good alternatives for certain tasks and certain types of artificial neural networks [105].

If $\boldsymbol{y} = (y_1, y_2, \dots, y_N)^T$ is a vector of the predicted values and $\bar{\boldsymbol{y}} = (\bar{y}_1, \bar{y}_2, \dots, \bar{y}_N)$ is the vector of the true values, then the L_2 loss function is usually defined as

$$E = ||\boldsymbol{y} - \bar{\boldsymbol{y}}||_2 = ||\bar{\boldsymbol{y}} - \boldsymbol{y}||_2 = \sum_{i=1}^{N} (\bar{y}_i - y_i)^2, \tag{8.31}$$

which is essentially the same as Eq. (8.13), except for a convenient factor of $1/2$. This can also be converted to the mean square error form as

$$E_m = \frac{1}{N} \sum_{i=1}^{N} (\bar{y}_i - y_i)^2. \tag{8.32}$$

In comparison with the smooth L_2-norm loss, the L_1-norm based loss function is given by

$$E_1 = ||\bar{\boldsymbol{y}} - \boldsymbol{y}||_1 = \sum_{i=1}^{N} |\bar{y}_i - y_i|, \tag{8.33}$$

which can also be interpreted as the mean absolute error (by multiplying a factor of $1/N$)

$$\frac{1}{N}\sum_{i=1}^{N} |\bar{y}_i - y_i|. \tag{8.34}$$

We have seen in earlier chapters that, for binary classification tasks, the cross entropy

$$E_c = -\sum_{i=1}^{N}\Big[\bar{y}_i \log y_i + (1-\bar{y}_i)\log(1-y_i)\Big] \tag{8.35}$$

can be used as a loss function, which is a measure of the differences between the true values and predicted values in the probabilistic sense. Thus a larger cross entropy means a larger difference.

The Kullback–Leibler (KL) divergence loss function is defined by

$$E_{KL} = \frac{1}{N}\sum_{i=1}^{N} \bar{y}_i \log(\bar{y}_i) - \frac{1}{N}\sum_{i=1}^{N} \bar{y}_i \log y_i, \tag{8.36}$$

which is a combination of the entropy (the first sum) and the cross entropy (the second term).

For other applications such as the classification using the support vector machines, the loss function can be defined as the hinge loss

$$E_h = \sum_{i=1}^{N} \max\{0, 1 - \bar{y}_i \cdot y_i\}, \tag{8.37}$$

which can also be defined with a scaling factor 1/2 as

$$E_h = \sum_{i=1}^{N} \max\{0, \frac{1}{2} - \bar{y}_i \cdot y_i\}. \tag{8.38}$$

Another related hinge loss function is the so-called squared hinge loss function

$$E_{h^2} = \sum_{i=1}^{N} \big(\max\{0, 1 - \bar{y}_i \cdot y_i\}\big)^2. \tag{8.39}$$

There are over a dozen other loss functions, including the Tanimoto loss, Chebyshev loss, and Cauchy–Schwarz divergence. For a detailed review of loss functions, we refer the readers to a more advanced literature such as [82] and [28].

8.5 Optimizers and choice of optimizers

As we have seen before, neural networks, especially multilayered networks, require the calculation of the gradient vectors iteratively. This can become very expensive if the numbers of weights and outputs are huge, which is true for deep learning. Therefore, some approximation or reduction in gradient calculation can speed up the learning process. Stochastic gradient descent (SGD) is one of such methods [19–21].

Briefly speaking, in the standard gradient descent, we need to calculate n values of the n components of the gradient vector ∇E, and n varies with layers. For example, $n = N$ for the output layer, whereas $n = m$ for the hidden layers. Instead of using the full gradient, the SGD uses a single observation or data example ∇E_i $(i = 1, 2, \ldots, n)$ to approximate ∇E, that is,

$$\boldsymbol{w}^{t+1} = \boldsymbol{w} - \eta \nabla E_i, \tag{8.40}$$

which can reduce the number of calculations by a factor of n. The choice of i can be the current data set. As this gradient instance is online and random, it is thus called the stochastic gradient or online gradient.

However, this extreme reduction may lead to inaccurate estimates of the true gradient. A better estimate can be obtained by a subset of stochastic gradient averaged. Alternatively, some extra term can be introduced:

$$\boldsymbol{w}^{t+1} = \boldsymbol{w} - \eta \nabla E_i + \tau \Delta \boldsymbol{w}, \tag{8.41}$$

where $0 \leq \tau \leq 1$ is the parameter controlling the inertia or momentum. This method is thus called the momentum method.

In Chapter 3 (Section 3.3), we have introduced most widely used optimizers for deep learning, including SGD and many momentum-based optimizers such as Adam and RMSprop.

An important issue for deep learning is the choice of a suitable optimizer. As we have seen in this book, the choice of optimization techniques may depend on the type of problem, solution quality, computational efforts, user's expertise, available resources, and other factors such as time constraints. In the context of neural networks and deep learning, it largely depends on the structure and depth of the neural networks, data type, size of the problem, and others.

The recent literature suggests that both Adam and RMSprop optimizers are among the best and can converge very fast. We refer the interested readers to a more advance literature such as the practical recommendations by Bengio [14].

8.6 Network architecture

The network structure of neural networks we have discussed is simple and easy to understand. Perceptrons are organized in a layered structure, and each layer is fed into the next layer, which forms a feed-forward structure. Each neuron in a layer can be fully connected to the next layer, but there is no connection between neurons in the same layer.

Feed-forward structure is just one popular structure; there are many other architectures that may have different advantages for different applications. There are comprehensive reviews on the architectures of neural networks. For example, Bengio [10] provided a comprehensive review on learning deep architectures for artificial intelligence, and van Veen [142] provided a detailed introductory review of various neural network architectures with colored topological representations.

Here we only briefly highlight a few different types of architecture and their main features.

- Feed-forward networks: the simplicity of the popular feed-forward neural networks allows the back propagation algorithm to work effectively. In the particular case where the input and output layers have identical structure and the outputs are trained to be the same as the inputs, this network becomes an autoencoder (AE). An interesting extension of the autoencoder is the denoising autoencoder (DAE), which allows both the input data and noise to feed into the network, and the trained outputs remove the noise. This can make the network more robust under noise or uncertainty [70,71]. In addition, if the inputs (and also outputs) for an autoencoder are sampling points from a probability distribution then such an AE network becomes a variational autoencoder (VAE), and its underlying foundation is variational Bayesian (VB) statistics that aims to optimize the posterior distributions. Furthermore, when different layered structures are organized in series, a deep neural net can be formed. Convolutionary neural network (CNN) is such a net, and we will introduce CNN later in this chapter.
- RBF networks: When the activation is carried out using a neighbor (in the manner similar to the k-nearest neighbor method), the activation function depends on the distance where radial basis functions (RBF) are used for activating each neuron, and then the feed-forward network becomes an RBF network.
- Recurrent neural network (RNN): This network architecture was developed by Elman [44], which uses a directed connection between every pair of neurons, and the connections are not just from the previous layer to the current layer, but also within the current layer itself. The connection strengths are modeled as time-varying real-valued weights. The training and minimization of errors are done using back propagation through time.
- Hopfield network: For the Hopfield network [77], each neuron is fully connected symmetrically with every other neuron in the network. There is a guarantee in terms of convergence for this network; however, it is not a recurrent network in general.
- Self-organized map: Strictly speaking, a self-organized map (SOM), or a Kohonen network [89], is an unsupervised learning technique that maps inputs onto outputs by activating most appropriate neurons and their neighbor neurons that can closely fit the inputs.
- Boltzmann machines: Boltzmann machines (BM) are a class of networks for machine learning, which has certain similarities to Kohonen networks. A particular

class of BMs is the so-called restricted Boltzmann machine (RBM), whose connections are restricted to feed-forward structures. We will introduce the RBM later in this chapter.

- Deep belief networks: A deep belief network (DBN) [13] is a deep multilayer neural network that has many levels of nonlinearities with a deep network architecture using restricted Boltzmann machines or variational autoencoders. The top-level prior is a restricted Boltzmann machine between layer $j-1$ and layer j, and some greedy layerwise training algorithm is often used to train one layer at a time. Then some fine-tuning of parameters of all layers is carried out.
- Modular neural networks: A modular neural network (MNN) uses a combination of a set of multiple neural networks, and each network is considered as a module. Each module can only carry out a specific task, and connection to other modules is designed on the system level. The training of the complex larger network can be subdivided into smaller modular training iteratively.
- Spiking neural networks: The spiking neural network (SNN) aims to provide a more realistic model for neural networks that are based on the synaptic structures and states [97]. It considers the time-varying information where signal impulses (a spiking train) rise for a short period and then gradually decay. The activation depends on the spiking impulse time interval, strength, and frequency. Such spiking networks can handle tasks that can be solved by classic neural networks and can also deal with more challenging tasks.

Other network architectures include neural Turing machine [56], generative adversarial nets (GAN) [55], and others such as extreme learning machines. Interested readers can consult a more advanced literature and textbooks.

8.7 Deep learning

Deep learning is not just a buzzword nowadays, but a very powerful tool for many applications related to artificial intelligence [27,105,126,134]. Image processing, pattern recognition, and speech recognition become much more accurate because of the effective use of deep learning techniques such as deep convolutionary neural networks (CNN) and restricted Boltzmann machines.

8.7.1 Convolutional neural networks

Deep learning involves neural networks with multiple hidden layers and usually has a modular structure of combing multiple convolutional neural networks (CNN) in a series of operations. A standard ANN is a fully connected network, and the computation of the next layer is carried out by matrix multiplication, whereas a CNN has sparse connections, and the computation is mainly done by convolution.

The main structure of a convolutional neural network (CNN) is that it has a convolution layer to convert inputs such as an image into a set of convoluted features by going through a filter or kernel of a fixed size. Then the convoluted features go

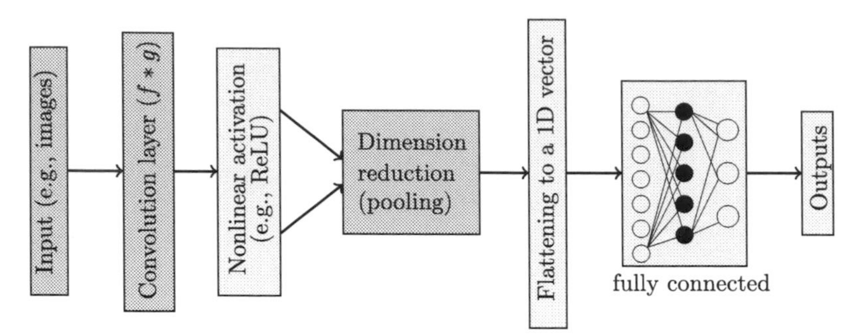

Figure 8.5 Schematic representation of the main CNN structure.

through activation and are then subsampled by pooling to reduce their dimensionality. Afterward, a pooled structure is flattened into a one-dimensional feature vector and is then fed into a fully or densely connected neural networks for classification. Obviously, multiple convolution layers with pooling can be used. The main structure can be summarized as the building blocks in Fig. 8.5.

8.7.1.1 Convolution and activation

The role of convolution is a crucial step because it allows us to use spare connections and focus on the local regions. If a simple 128×128 grayscale image is fed into a fully connected neural network, then the inputs can be represented by a one-dimensional vector with $128 \times 128 = 16384$ elements. Thus this input vector may require 16384 connections to each neuron in the next layer, which means that it needs 16384 weights or parameters to define such connections for this neuron. For large images, the number of free parameters can be astronomical, and thus a fully connected neural network is not practical for such applications. One way is using a kernel with few parameters, and such a kernel should be able to apply for the larger image with the same set of parameters. Such kernel-based operations are essentially convolutions.

The basic idea of convolution between a function $f(x)$ and a kernel h is defined as

$$(f * h)(x) = \int_{-\infty}^{+\infty} f(x-\tau)h(\tau)d\tau = \int_{-\infty}^{+\infty} f(\tau)h(x-\tau)d\tau, \tag{8.42}$$

which is the integral of the overlapped area between two functions. This is equivalent to sliding one function over the other function. In the simple case where h is a simple unit slot sliding along a function $f(x)$, the convolution value at any x is the overlapped shaded area (see Fig. 8.6).

Convolution has been used extensively in signal and image processing. In the case of two-dimensional (2D) inputs such as images $f(x, y)$, the 2D convolution can be

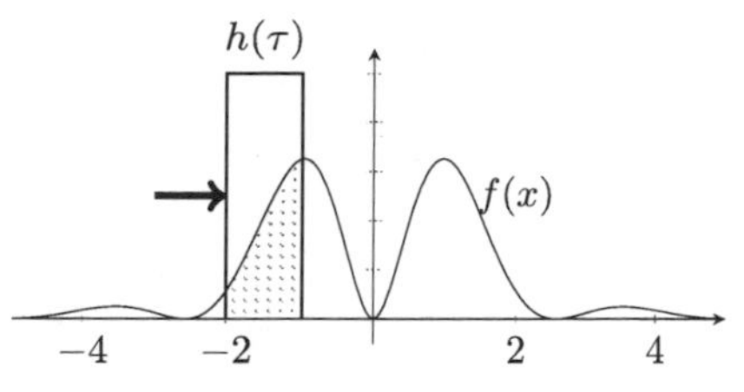

Figure 8.6 Simple convolution of $f(x)$ and kernel $h(\tau)$.

written as

$$(f * h)(x, y) = \int_{-\infty}^{+\infty} \int_{-\infty}^{+\infty} h(p, q) f(x - p, y - q) dp dq$$
$$= \int_{-\infty}^{+\infty} \int_{-\infty}^{+\infty} f(p, q) h(x - p, y - q) dp dq, \quad (8.43)$$

which usually becomes a double sum over a region of an image. Here the kernel $h(x, y)$ acts as a filter or operator. For example, a simple edge detection kernel is

$$h = \begin{pmatrix} +1 & 0 & -1 \\ 0 & 0 & 0 \\ -1 & 0 & +1 \end{pmatrix}, \quad (8.44)$$

which is a differential operator along the x and y directions. Another edge detection operator is

$$g = \begin{pmatrix} 0 & 1 & 0 \\ 1 & -4 & 1 \\ 0 & 1 & 0 \end{pmatrix}, \quad (8.45)$$

which is a gradient-based kernel.

The convolution action of a kernel on a 2D image essentially mimics the receptive field of the biological sensory neural systems. The intensity of a grayscale image is represented by a 2D array, and each pixel has a value between 0 to 255. However, for simplicity, we assume that the values are between 0 to 9 and apply a simple kernel h of 3×3 as an edge detector to a small image of 6×6 (see Fig. 8.7). The convolution of h with f gives a 4×4 matrix W. The convolution of h with the first 3×3 region (top left corner, dashed) gives

$$W(1, 1) = (f * h) = +1 \times 8 + 0 \times 0 + (-1) \times 2$$
$$+ 0 \times 5 + 0 \times 7 + 0 \times 9 + (-1) \times 2 + 0 \times 4 + 1 \times 6 = 10. \quad (8.46)$$

Similarly, the convolution of h with the second region (shown as dotted) is

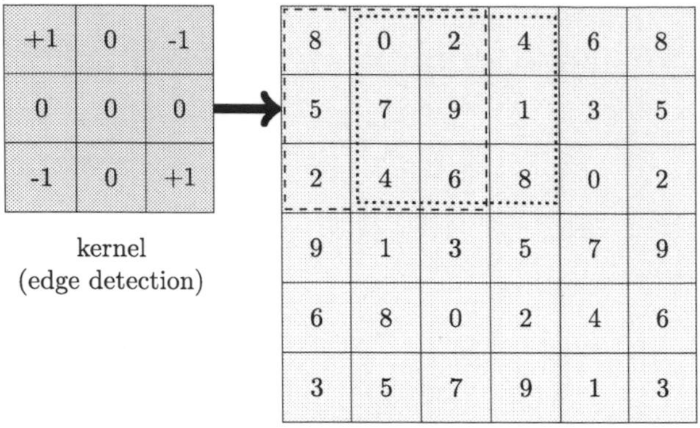

+1	0	-1
0	0	0
-1	0	+1

kernel
(edge detection)

8	0	2	4	6	8
5	7	9	1	3	5
2	4	6	8	0	2
9	1	3	5	7	9
6	8	0	2	4	6
3	5	7	9	1	3

Figure 8.7 The schematic representation of a 2D convolution.

$$W(1,2) = 1 \times 0 + 0 \times 2 + (-1) \times 4$$
$$+ 0 \times 7 + 0 \times 9 + 0 \times 1 + (-1) \times 4 + 0 \times 6 + 1 \times 8 = 0. \quad (8.47)$$

The rest of W entries can be calculated in a similar manner.

It is worth pointing out that the convolution result W has a size of 4×4 for an image of 6×6, and this is the case without padding. Sometimes, padding is used to make sure that the resulting matrix has the same size. In general, without padding, the convolution matrix has a size of $[n - (k+1)/2] \times [n - (k+1)/2]$ for an image of $n \times n$ with a kernel of $k \times k$. In the previous simple example, we have $n = 6$ and $k = 3$.

The previous convolution has been demonstrated for a grayscale image. For a colored image, it is usually represented as a three-dimensional array for three different color channels (Red, Green, and Blue). This means that the convolution kernel should be applied for each color channel, and thus the resulting convolution matrix is a 3D array or rank-3 tensor in the most general sense. We refer the interested readers to the more advance literature [90,93].

After the convolution operations, the outputs can fed into nonlinear activation. Activation is usually carried out by a rectified linear unit (ReLU) (essentially $\max(0, x)$) and Leaky ReLU as discussed earlier in this chapter.

In general, a convolution block consists of a convolution layer and an activation layer. A convolutional neural network can have multiple convolution blocks before sending to dimension reduction via pooling.

8.7.1.2 Pooling

Pooling is a dimensional reduction technique by focusing a region of size $k \times k$ as a pooling filter. Only one value is calculated from a region of fixed size, and this value is usually taken as the maximum value or average of all the values. In the case of using the maximum value, the pooling operation is called the max pooling. For example, in the simple pooling shown in Fig. 8.8, a 2×2 pooling filter is applied to every

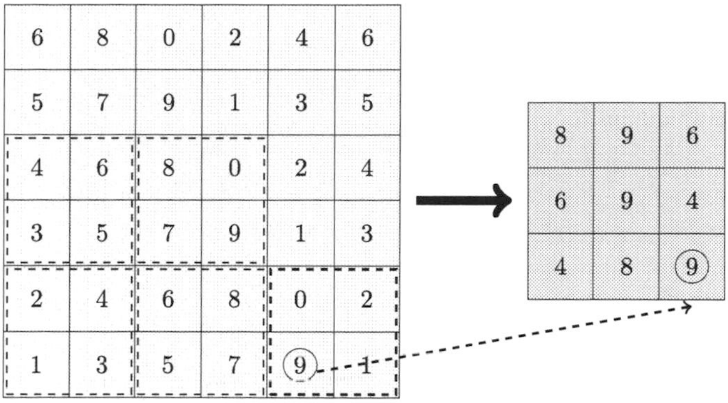

6	8	0	2	4	6
5	7	9	1	3	5
4	6	8	0	2	4
3	5	7	9	1	3
2	4	6	8	0	2
1	3	5	7	9	1

8	9	6
6	9	4
4	8	9

Figure 8.8 Max pooling in 2D with a 2×2 pooling filter.

nonoverlapping 2×2 subregions, and the bottom right corner has

$$\begin{vmatrix} 0 & \vdots & 2 \\ \cdots & \cdots & \cdots \\ 9 & \vdots & 1 \end{vmatrix}, \tag{8.18}$$

whose maximum value is 9, which gives a single value 9 as shown inside a circle.

The size reduction can be significant, and in most cases, the size can be reduced at least by half. In case of the stride, it is also k, which means that no two regions are overlapping, and the final pooled size is $n/k \times n/k$ for an image of $n \times n$ with a pooling filter of $k \times k$. Obviously, n/k should be rounded to the nearest integers. For example, when $n = 100$ and $k = 2$, the original size of 100×100 can be reduced to 50×50, which is a quarter of the original size. A simple case of 2×2 pooling is shown in Fig. 8.8.

It is worth pointing out that both ReLU and max pooling use "max", and thus the order of ReLU and max pooling can be interchangeable. The results should be same if either ReLU or max pooling is the first. This property is sometimes called equivariance. However, in most deep learning packages such as TensorFlow, ReLU is usually done first.

The equivariant representation is one useful property of CNNs. For example, for an image, convolution, and translation can be interchangeable, that is, the results should be the same if convolution is done first (then translation) or translation first (followed by convolution). Loosely speaking, two mathematical operations ϕ and ψ are said to be equivariant if $\phi(\psi(x)) = \psi(\phi(x))$.

8.7.1.3 Flattening

The output from convolution and pooling operations are either 2D arrays for grayscale images or 3D arrays for color images. In order to feed such outputs for further learning by feed-forward neural networks, it is more convenient to reshape such 2D and 3D arrays into a one-dimensional feature vector. This process is called flattening, which can be done by stacking each row or column in a sequential order, and then at different depths. In general, for 2D arrays of size $n \times n$, the flattened 1D vector has n^2 elements, for example, for gray images of size $n \times n \times 1$.

If the convolution kernel has a size of 3×3 and the number of hidden neurons is $m = 16$, then the convoluted features have a size of $(n-2) \times (n-2) \times m$ without padding. If a pooling is done over a $k \times k$ grid with a stride k (nonoverlapping), then pooled features become $(n-2)/k \times (n-2)/k \times m$. In case of $n = 32$, $m = 16$, and $k = 2$, the pooled features form a 3D array of size $15 \times 15 \times 16$. This can be further converted to a one-dimensional array by flattening into a feature vector with $15 \times 15 \times 16 = 3600$ elements or features.

For the 2×2 max pooling shown in Fig. 8.8, the 2D array on the right can be flattened to

$$\begin{pmatrix} 8 & 9 & 6 \\ 6 & 9 & 4 \\ 4 & 8 & 9 \end{pmatrix} \Longrightarrow \begin{pmatrix} 8 & 9 & 6 & 6 & 9 & 4 & 4 & 8 & 9 \end{pmatrix}^T. \tag{8.49}$$

8.7.1.4 Fully connected neural network

It is worth pointing out that the output features from a convolution layer are low-dimensional regional features. To obtain high-level nonlinear features that are appropriate for classification and recognition, a fully connected neural network should be used.

From the above, the flattened one-dimensional feature vector can be fed into a fully connected neural network with hidden layers. The final training outputs put into different categories using softmax and any appropriate probabilistic interpretations. The main role of softmax operations is to convert a real-valued vector to a normalized form as a probability distribution, which assigns a probability to each possible class and thus allows the ease of interpretation of classification and categorization. The overall architecture of deep neural networks can be very sophisticated, and different software packages may use different combinations of different building blocks. The well-known AlexNet/ImageNet [90] and Google's TensorFlow [59] use different architectural structures.

In addition, neural networks can have some dropout operations. The main idea of dropping out is that neurons in an ANN can be ignored and removed with probability p, which can reduce the number of parameters and thus reduce the complexity of the network. This can also potentially prevent overfitting [135,94]. Typically, the probability $p = 0.25$ is often used.

Furthermore, the values for image processing are between 0 to 255. In case of other inputs, in some applications, the ranges of different signals can be very different, and some normalization is usually needed to scale the inputs to about the same or comparable ranges.

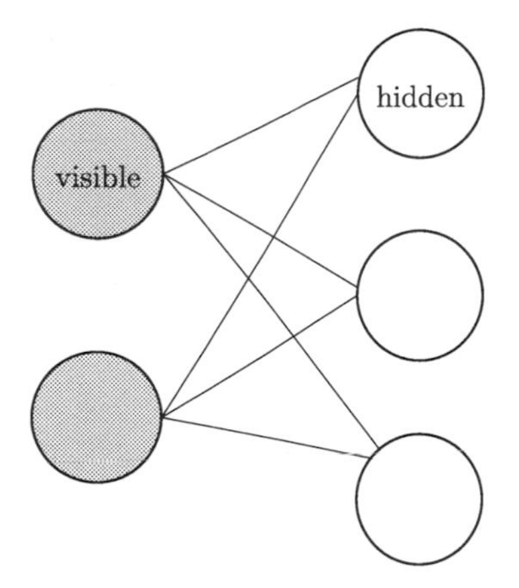

Figure 8.9 Restricted Boltzmann machine with visible and hidden units.

8.7.2 Restricted Boltzmann machine

A very useful tool for deep learning applications is the restricted Boltzmann machine (RBM), which is a two-layer (or two-group) Boltzmann machine with m visible units v_i $(i = 1, 2, \ldots, m)$ and n hidden units h_j $(j = 1, 2, \ldots, n)$ where both v_i and h_j are binary states. The visible unit i has a bias α_i, and the hidden unit j has a bias β_j, whereas the weight connecting them is denoted by w_{ij}. The restriction is that their neuron units connecting visible and hidden units form a bipartite graph (see Fig. 8.9), whereas no connection with the same group (visible or hidden) is allowed.

Using the notations and configuration given in 2010 by Hinton [72], a restricted Boltzmann machine can be represented by a pair $(\boldsymbol{v}, \boldsymbol{h})$ where $\boldsymbol{v} = (v_1, v_2, \ldots, v_m)^T$ and $\boldsymbol{h} = (h_1, h_2, \ldots, h_n)^T$. The energy of the system can be calculated by

$$E(\boldsymbol{v}, \boldsymbol{h}) = -\sum_{i \in \text{visible}} \alpha_i v_i - \sum_{j \in \text{hidden}} \beta_j h_j - \sum_{i,j} w_{ij} v_i h_j. \tag{8.50}$$

The probability of a network associated with every possible pair of a visible vector $\boldsymbol{v}$ and a hidden vector $\boldsymbol{h}$ is assumed to obey the Boltzmann distribution

$$p(\boldsymbol{v}, \boldsymbol{h}) = \frac{1}{Z} e^{-E(\boldsymbol{v}, \boldsymbol{h})}, \tag{8.51}$$

where Z is a normalization constant, also called the partition function, which is essentially the summation over all possible configurations, that is,

$$Z = \sum_{\boldsymbol{v}, \boldsymbol{h}} e^{-E(\boldsymbol{v}, \boldsymbol{h})}. \tag{8.52}$$

The marginal probability of a network associated with $\boldsymbol{v}$ can be calculated by the sum over all possible hidden vectors in (8.51), so we have

$$p(\boldsymbol{v}) = \frac{1}{Z} \sum_{\boldsymbol{h}} e^{-E(\boldsymbol{v},\boldsymbol{h})}. \tag{8.53}$$

The essential idea of using RBM for training over a set of data (such as images) is to adjust the weights and bias values so that a training image can maximize its associated network probability (thus minimizing the corresponding energy) [46,92]. For a training set, the maximization of the joint probability $p(\boldsymbol{v})$ is equivalent to the maximization of the expected log probability (i.e., $\log p(\boldsymbol{v})$). Since

$$\frac{\partial \log p(\boldsymbol{v})}{\partial w_{ij}} = <v_i h_j>_{\rm data} - <v_i h_j>_{\rm RBMmodel}, \tag{8.54}$$

we can calculate the adjustments in weights by using the stochastic gradient method

$$\Delta w_{ij} = \eta\big(<v_i h_j>_{\rm data} - <v_i h_j>_{\rm RBMmodel} \big), \tag{8.55}$$

where $<v_i h_j>$ means the expectation over the associated distributions. It is worth pointing out that the stochastic gradient ascent (in contrast with the stochastic gradient descent) is used.

The individual activation probabilities for visible and hidden units are determined by the sigmoid function $S(x) = 1/(1+e^{-x})$, that is,

$$p(v_i = 1|\boldsymbol{h}) = S\Big(\alpha_i + \sum_j w_{ij} h_j\Big) \tag{8.56}$$

and

$$p(h_j = 1|\boldsymbol{v}) = S\Big(\beta_j + \sum_i w_{ij} v_i\Big). \tag{8.57}$$

The restricted Boltzmann machines form the essential part of deep belief networks with stacked RBM layers.

There are many good software packages for artificial neural networks, and there are dozens of good books fully dedicated to theory and implementations. Therefore, we will not provide any code here.

8.7.3 Deep neural nets

The essence of deep learning is the feedforward deep neural network (i.e., multilayer perceptrons) with different levels of abstraction, and training of such multilayer perceptron networks (MLPN) is done by the backpropagation algorithm. Here "deep" means multiple layers, which can range from a few layers to several hundred layers or even more. Such learning can be done by one layer at a time. Deep learning intends to learn representations and feature hierarchies from higher-level features or complicated concepts formed by lower-level features or simpler concepts, and such a system consists of multiple levels of abstraction without human-crafted feature creation [105,67].

In order for the deep networks to work well, it requires both a large amount of training data (with a wide range of diversity) and computing power to realize the training practically. For example, for face recognition and image classifications, the number of images can vary from a few million to hundreds of millions. Therefore, it is no surprise that big companies such as Google, Microsoft, IBM, and Amazon are among those having the most successful deep learning systems. AlexNet and TensorFlow are good examples of deep learning software packages. There are also many good tutorials on these vast topics. For example, the Stanford website udldl.stanford.edu/wiki/index.php can be a very helpful starting point.

The literature about deep learning, especially convolutional neural networks, is vast. We refer the interested readers to more advanced and specialized journal papers and books [68,69,54,95].

8.7.4 Trends in deep learning

Deep learning is a very active research area, and new progress is being made every year or even every week. Here we only briefly outline some recent developments and trends in this area [54,71,104]. The intention is not for completeness, but rather to inspire further research.

- Self-taught teaching is a two-stage learning approach where learning is carried out first on the unlabeled data to learn a representation, and then this learned representation is applied to labeled data for classification tasks [119]. There is no assumption that the labeled data were drawn from the same distributions for the unlabeled data, and thus the data can be in different classes and from different distributions. In other words, self-taught learning can be considered as transfer learning from unlabeled data, or unsupervised transfer.
- Transfer learning transfers knowledge from one supervised learning task to another, which requires additional labeled data from a different (but related) task [114,11,141]. The requirement of such extra labeling may be expensive.
- One-shot learning is an efficient learning technique where learning is carried out with only one example or handful examples [45]. Instead of learning from scratch, training is improved from previously learned categories, which is in general in terms of Bayesian inference. Both one-shot learning and transfer learning can belong to augmented and lean learning paradigm.
- Generative adversarial networks (GAN) use one network to generate candidates such as realistic-looking images and other network to learn and detect if the generated candidates are real or not. The generated samples can be sufficient realistic to the real things [55]. Bayesian GANs can implicitly learn rich distribution over data such as image and audio [55], and such data may be difficult to model with an explicit likelihood probability. Thus Bayesian GAN can be considered as a Bayesian approach for unsupervised and semisupervised learning with GANs, in combi-

nation with a stochastic gradient-based Monte Carlo. In essence, it is a hybrid learning model.

- Capsule networks are a new network architecture with a hierarchical spatial relationship where a capsule is a group of neurons whose activity vector represents certain entity parameters [128,73]. The entity can be an object or part of an image. Active capsules at one-level make predictions by using transformation matrices, and a higher-level capsule becomes active if multiple predictions agree. Sabour et al. [128] showed that a trained multilayer capsule system can achieve state-of-the-art performance or better performance.

There are other developments and architectures, for example, the combination of recurrent neural networks (RNN) with CNN, long short-term memory (LSTM) networks, the RNN with long short-term memory, gated recurrent neural networks, region proposal networks, deep reinforcement learning, and others. We will not have any space here to introduce them briefly.

The rapid developments of deep learning has made it possible to apply artificial intelligence in many applications such as image classification and captioning, text generation, natural language processing and translation, big data science, product recommendation, and many others.

Interested readers can consult a more advanced literature on these topics.

8.8 Tuning of hyperparameters

Many techniques we have discussed in this chapter and previous chapters have algorithm-dependent hyperparameters. For example, k in the k-means, λ in the regularization methods, the learning rate η, and the number of layers in deep nets are all hyperparameters. At the moment, the choices of such hyperparameters are mainly by researchers, based on parametric studies, the type of problems, and empirical studies, even personal expertise and experience. The initial choice can be an educated guess, then fine-tuning is attempted. Some researchers use uniform grid-style search, whereas others use various approaches. In fact, the optimal choice of such hyperparameters is a challenging task, it is the optimization of an optimization problem, and the choice is largely problem-dependent.

In the context of deep learning, the learning rate can be given an initial rate, and then a monotonic decay schedule is used to gradually reduce the effective learning rates. Some researchers prefer to use randomized learning rates [15], and others prefer to use the Bayesian approach to set hyperparameters.

Currently, there still lacks rigorous theory for choosing such parameters. Recent studies suggested random search, heuristic methods, and metaheuristic algorithms such as genetic algorithms, and the firefly algorithm can be very useful to optimize hyperparameters. Interested readers can consult a more advance literature such as some journal articles on these topics.

8.9 Notes on software

There are many software packages on ANN, CNN, and deep learning. For example, Matlab has the `nntool`, `alexnet`, and `Googlenet`. R has toolboxes such as `elasticnet`, deep learning `deepnet`, Restricted Boltzmann machine `RcppDL`, and an interface to TensorFlow.

Python has machine learning packages such as `scikit-learn`, including support vector machines and neural networks.

Google's TensorFlow is a very powerful engine for deep learning. There are many frontend tools to interact with TensorFlow. For example, both `Keras` are `TFLearn` arc among the most widely used, and they are easy to start with many examples on the Internet.

Other tools include `Theano`, `Torch`, `Caffe`, and others. In addition, Microsoft, IBM, Amazon, and many other companies have deep learning tools. The sources are diverse, and the literature is vast.

Bibliography

[1] C.C. Aggarwal, Data Mining: The Textbook, Springer, New York, 2015.
[2] H. Akaike, A new look at the statistical model identification, IEEE Transactions on Automatic Control 19 (6) (1974) 716–723.
[3] J. Aldrich, R. A. Fisher and the making of maximum likelihood 1912–1922, Statistical Science 12 (3) (1997) 162–176.
[4] E. Alpaydin, Introduction to Machine Learning, MIT Press, Cambridge, MA, 2004.
[5] A. Antoniou, W.S. Lu, Practical Optimization: Algorithms and Engineering Applications, Springer, 2007.
[6] S. Arara, B. Barak, Computational Complexity: A Modern Approach, Cambridge University Press, Cambridge, 2009.
[7] M. Bartholomew-Biggs, Nonlinear Optimization With Engineering Applications, Springer, 2008.
[8] D. Basak, S. Pal, D.C. Patranabis, Support vector regression, Neural Information Processing 11 (10) (2007) 203–224.
[9] I. Ben-Gal, Bayesian networks, in: F. Ruggeri, F. Faltin, R. Kenett (Eds.), Encyclopedia of Statistics in Quality and Reliability, John Wiley & Sons, Hoboken, NJ, 2007.
[10] Y. Bengio, Learning deep architecture for AI, Foundations and Trends in Machine Learning 2 (1) (2009) 1–127.
[11] Y. Bengio, Deep learning of representations for unsupervised and transfer learning, in: I. Guyon, G. Dror, V. Lemaire, G. Taylor, D. Silver (Eds.), JMLR Workshop and Conference Proceedings, Workshop on Unsupervised and Transfer Learning, vol. 27, 2012, pp. 17–37.
[12] Y. Bengio, N. Boulanger-Lewandowski, R. Pascanu, Advances in optimizing recurrent networks, preprint, arXiv:1212.0901 [cs.LG], 2012. (Accessed 10 August 2018).
[13] Y. Bengio, P. Lamblin, D. Popovici, H. Larochelle, Greedy layer-wise training of deep networks, Advances in Neural Information Processing Systems 19 (2006) 153–160.
[14] Y. Bengio, Practical recommendations for gradient-based training of deep architectures, in: G. Montavon, G.B. Orr, K.R. Müller (Eds.), Neural Networks: Tricks of the Trade, in: Lecture Notes in Computer Science, vol. 7700, 2012, pp. 437–478.
[15] J. Bergstra, Y. Bengio, Random search for hyper-parameter optimization, Journal of Machine Learning Research 13 (2012) 281–305.
[16] D. Bertsekas, R. Gallager, Data Networks, 2nd edition, Prentice Hall, Englewood Cliffs, NJ, 1992.
[17] C. Bishop, Pattern Recognition and Machine Learning, Springer, New York, 2007.
[18] L. Bottou, Online algorithms and stochastic approximations, in: Online Learning and Neural Networks, Cambridge University Press, Cambridge, 1998.
[19] L. Bottou, Stochastic learning, in: O. Bousquet, U. von Luxburg (Eds.), Advanced Lectures on Machine Learning, in: Lecture Notes in Artificial Intelligence, Springer-Verlag, Berlin, 2004.
[20] L. Bottou, Large-scale machine learning with stochastic gradient descent, in: Y. Lechevallier, G. Saporta (Eds.), Proceedings of COMPSTAT'2010, 2010, pp. 177–186.

[21] L. Bottou, Stochastic gradient descent tricks, in: G. Montavon, G.B. Orr, K.R. Müller (Eds.), Neural Networks: Tricks of the Trade, Reloaded Edition, in: Lecture Notes in Computer Science, vol. 7700, 2012, pp. 1–16.
[22] S.P. Boyd, L. Vandenberghe, Convex Optimization, Cambridge University Press, Cambridge, 2004.
[23] L. Breiman, Random forests, Machine Learning 45 (2001) 5–32.
[24] J.S. Bridle, Probabilistic interpretation of feedforward classification network outputs, with relationships to statistical pattern recognition, in: F.F. Soulié, J. Hérault (Eds.), Neurocomputing: Algorithms, Architectures and Applications, in: NATO ASI Series F: Computer and System Sciences, vol. 68, Springer, Berlin, 1989, pp. 227–236.
[25] P.S. Bradley, U.M. Fayyad, C. Reina, Scaling clustering algorithms to large databases, in: Proc. Knowledge Discovery and Data Mining, 1998, pp. 9–15.
[26] J.F. Box, R.A. Fister, The Life of a Scientist, John Wiley & Sons, New York, 1978.
[27] N. Buduma, Fundamentals of Deep Learning: Designing Next-Generation Artificial Intelligence Algorithms, O'Reilly Media, Sebstopol, CA, 2015.
[28] I. Changhau, Loss functions in neural networks, Github online article, 7 June 2017, https://isaacchanghau.github.io/post/loss_functions/. (Accessed 10 November 2018).
[29] E. Charniak, Bayesian networks without tears, AI Magazine 12 (4) (1991) 50–63.
[30] B.A. Cipra, The best of the 20th century: editors name top 10 algorithms, SIAM News 33 (4) (2000) 1–2.
[31] P. Comon, Independent component analysis – a new concept?, Signal Processing 36 (1994) 287–314.
[32] A.R. Conn, N.I.M. Gould, P.L. Toint, Trust-Region Methods, SIAM & MPS, 2000.
[33] C. Cortes, V. Vapnik, Support-vector networks, Machine Learning 20 (3) (1995) 273–297.
[34] N. Cristianini, J. Shawe-Taylor, An Introduction to Support Vector Machines and Other Kernel-Based Learning Methods, Cambridge University Press, Cambridge, 2000.
[35] G.B. Dantzig, M.N. Thapa, Linear Programming 1: Introduction, Springer-Verlag, New York, 1997.
[36] A.P. Dempster, N.M. Laird, D.B. Rubin, Maximum likelihood from incomplete data via the EM algorithm, Journal of the Royal Statistical Society, Series B 39 (1) (1977) 1–38.
[37] C. Dhaenens, L. Jourdan, Metaheuristics for Big Data, John Wiley & Sons, Hoboken, NJ, 2016.
[38] N.R. Draper, H. Smith, Applied Regression Analysis, 3rd edition, John Wiley & Sons, New York, 1998.
[39] B.A. Draper, K. Baek, M.S. Bartlett, J.R. Beveridge, Recognizing faces with PCA and ICA, Computer Vision and Image Understanding 91 (2003) 115–137.
[40] H. Drucker, C.J.C. Burges, L. Kaufman, A.J. Smola, V.N. Vapnik, Support vector regression machines, in: Advances in Neural Information Processing Systems, NIPS 1996, MIT Press, 1997, pp. 155–161.
[41] J. Duchi, E. Hazan, Y. Singer, Adaptive subgradient methods for online learning and stochastic optimization, Journal of Machine Learning Research 12 (2011) 2121–2159.
[42] S.Y. Elhabian, A. Farag, A Tutorial on Data Reduction: Linear Discriminant Analysis, Technical Report, Computer Vision and Image Processing Laboratory (CVIP Lab), University of Louisville, Sept 2009.
[43] K. Eriksson, D. Estep, C. Johnson, Applied Mathematics: Body and Soul, Volume 1: Derivatives and Geometry in IR3, Springer-Verlag, Berlin, 2004.
[44] J.L. Elman, Fining structure in time, Cognitive Science 14 (2) (1990) 179–211.
[45] L. Fei-Fei, R. Fergus, P. Perona, One-shot learning of object categories, IEEE Transactions on Pattern Analysis and Machine Intelligence 28 (4) (2006) 594–611.

[46] A. Fischer, C. Igel, Training restricted Boltzmann machines: an introduction, Pattern Recognition 47 (1) (2014) 25–39.

[47] R.A. Fisher, The use of multiple measurements in taxonomic problems, Annual of Eugenics 7 (2) (1936) 179–188.

[48] G.S. Fishman, Monte Carlo: Concepts, Algorithms and Applications, Springer, New York, 1995.

[49] R. Fletcher, Practical Methods of Optimization, 2nd edition, John Wiley & Sons, 2000.

[50] D.A. Freedman, Statistical Models: Theory and Practice, Cambridge University Press, Cambridge, 2009.

[51] A. Gelman, J.B. Carlin, H.S. Stern, D.B. Dunson, A. Vehtari, D.B. Rubin, Bayesian Data Analysis, 3rd edition, CRC Press, Boca Raton, FL, 2013.

[52] S. Geman, D. Geman, Stochastic relaxation, Gibbs distribution and Bayesian restoration of images, IEEE Transactions on Pattern Analysis and Machine Intelligence 6 (1984) 721–741.

[53] P.E. Gill, W. Murray, M.H. Wright, Practical Optimization, Emerald Group Publishing Ltd., Bingley, 1982.

[54] I. Goodfellow, Y. Bengio, A. Courville, Deep Learning, The MIT Press, Cambridge, MA, 2017.

[55] I.J. Goodfellow, J. Pouget-Abadie, M. Mirza, B. Xu, D. Warde-Farley, S. Ozair, A. Courville, Y. Bengio, Generative adversarial nets, arXiv:1406.2661, 2014. (Accessed 14 December 2018).

[56] A. Graves, G. Wayne, I. Danihelka, Neural Turing Machines, arXiv:1410.5401, 2014. (Accessed 15 December 2018).

[57] D.E. Goldberg, Genetic Algorithms in Search, Optimisation and Machine Learning, Addison Wesley, Reading, MA, 1989.

[58] O. Goldreich, Computational Complexity: A Conceptual Perspective, Cambridge University Press, Cambridge, 2008.

[59] Google tensor flow, https://www.tensorflow.org. (Accessed 1 November 2018).

[60] S. Guha, R. Rastogi, K. Shim, CURE: an efficient clustering algorithm for large databases, in: Proc. ACM SIGMOD Int. Conf. on Management of Data, vol. 27(2), 1998, pp. 73–84.

[61] D.J. Hand, K. Yu, Idiot's Bayes – not so stupid after all?, International Statistical Review 69 (3) (2001) 385–399.

[62] T. Hastie, R. Tibshirani, J. Friedman, The Elements of Statistical Learning: Data Mining, Inference, and Prediction, second edition, Springer Series in Statistics, Springer, Heidelberg, 2009.

[63] W.K. Hastings, Monte Carlo sampling methods using Markov chains and their applications, Biometrika 57 (1970) 97–109.

[64] S.S. Haykin, Neural Networks: A Comprehensive Foundation, Prentice Hall, New Jersey, 1999.

[65] J. Heaton, Artificial Intelligence for Humans, Volume 3: Deep Learning and Neural Networks, CreateSpace Independent Publishing, 2015.

[66] M. Hilbert, Big data for development: a review of promises and challenges, Development Policy Review 34 (1) (2016) 135–174.

[67] G.E. Hinton, S. Osindero, Y.W. Teh, A fast learning algorithm for deep belief nets, Neural Computation 18 (7) (2006) 1527–1554.

[68] G.E. Hinton, Learning multiple layers of representation, Trends in Cognitive Sciences 11 (2007) 428–434.

[69] G.E. Hinton, L. Deng, D. Yu, G. Dahl, A. Mohamed, N. Jaitly, A. Senior, V. Vanhoucke, P. Nguyen, T. Sainath, B. Kingsbury, Deep neural networks for acoustic modeling in speech recognition – the shared views of four research groups, IEEE Signal Processing Magazine 29 (6) (2012) 82–97.

[70] G.E. Hinton, R. Salakhutdinov, Reducing the dimensionality of data with neural networks, Science 313 (5786) (2006) 504–507.

[71] G.E. Hinton, Deep belief networks, Scholarpedia 4 (5) (2009) 5947.

[72] G.E. Hinton, A Practical Guide to Training Restricted Boltzmann Machines, UTML TR 2010-003, Technical Report, University of Toronto, 2010.

[73] G.E. Hinton, A. Krizhevsky, S.D. Wang, Transforming auto-encoder, in: International Conference on Artificial Neural Networks, Springer, 2011, pp. 44–51.

[74] T.K. Ho, Random decision forests, in: Proceedings of the Third International Conference on Document Analysis and Recognition, Montreal, QC, 14–16 Aug 1995, 2016, pp. 278–282.

[75] J. Holland, Adaptation in Natural and Artificial Systems, University of Michigan Press, Ann Arbor, USA, 1975.

[76] D.E. Holmes, Big Data: A Very Short Introduction, Oxford University Press, Oxford, 2017.

[77] J.J. Hopfield, Neural networks and physical systems with emergent collective computational abilities, Proceedings of the National Academy of Sciences 79 (8) (1982) 2554–2558.

[78] J. Hurwitz, A. Nugent, F. Halper, M. Kaufman, Big Data for Dummies, John Wiley & Sons, Hoboken, NJ, 2013.

[79] A. Hyvärinen, E. Oja, Independent component analysis: algorithms and applications, Neural Networks 13 (4–5) (2000) 411–430.

[80] A. Hyärinen, Independent component analysis in the presence of Gaussian noise by maximizing joint likelihood, Neurocomputing 22 (1) (1998) 49–67.

[81] J.E. Jackson, A User's Guide to Principal Components, John Wiley & Sons, New York, 1991.

[82] K. Janocha, W.M. Czarnecki, On loss functions for deep neural networks in classification, arXiv:1702.05659, 2017. (Accessed 10 October 2018).

[83] I.T. Jolliffe, Principal Component Analysis, 2nd edition, Springer, New York, 2002.

[84] M.I. Jordan, Learning in Graphical Models, MIT Press, Cambridge, MA, 1999.

[85] C. Jutten, J. Hérault, Blind separation of sources, part I: an adaptive algorithm based on neuromimetic architecture, Signal Processing 24 (1) (1991) 1–10.

[86] V. Kecman, Learning and Soft Computing – Support Vector Machines, Neural Networks, Fuzzy Logic Systems, The MIT Press, Cambridge, MA, 2001.

[87] J. Kennedy, R.C. Eberhart, Particle swarm optimization, in: Proc. of IEEE International Conference on Neural Networks, Piscataway, NJ, 1985, pp. 1942–1948.

[88] D.P. Kingma, J. Ba Adam, A method for stochastic optimization, in: International Conference on Learning Representations, 2015, pp. 1–13. Also at arXiv:1412.6980 [cs.LG], December 2014. (Accessed 10 August 2018).

[89] T. Kohonen, Self-organized formation of topologically correct feature maps, Biological Cybernetics 43 (1) (1982) 59–69.

[90] A. Krizhevsky, I. Sutskever, G.E. Hinton, ImageNet classification with deep convolutional neural networks, Communications of the ACM 60 (6) (2017) 84–90.

[91] J. Lampinen, A. Vehtari, Bayesian approach for neural networks – review and case studies, Neural Networks 14 (1) (2001) 7–24.

[92] H. Larochelle, Y. Bengio, Classification using discriminative restricted Boltzmann machines, in: Proceedings of the 25th Int. Conference on Machine Learning, ICML2008, 2008, p. 536.
[93] Y. LeCun, Y. Bengio, G.E. Hinton, Deep learning, Nature 521 (2015) 436–444.
[94] Y. LeCun, L. Bottou, G.B. Orr, K.R. Müller, Efficient backdrop, in: G.B. Orr, K.R. Müller (Eds.), Neural Networks: Tricks of the Trade, in: Lecture Notes in Computer Science, vol. 1524, Springer, Berlin, 2012, pp. 9–48.
[95] Y. LeCun, L. Bottou, Y. Bengio, P. Haffner, Gradient-based learning applied to document recognition, Proceedings of the IEEE 86 (1998) 2278–2324.
[96] J. Leskovec, A. Rajaraman, J.D. Ullman, Mining of Massive Datasets, 2nd edition, Cambridge University Press, Cambridge, UK, 2014.
[97] W. Maass, Networks of spiking neurons: the third generation of neural network models, Neural Networks 10 (9) (1997) 1659–1671.
[98] D.J.C. MacKay, Information Theory, Inference, and Learning Algorithms, Cambridge University Press, Cambridge, UK, 2003.
[99] B. Marr, Big Data: Using Smart Big Data, Analytics and Metrics to Make Better Decisions and Improve Performance, John Wiley & Sons, Hoboken, NJ, 2015.
[100] A.M. Martínez, A.C. Kak, PCA versus LDA, IEEE Transactions on Pattern Analysis and Machine Intelligence 23 (2) (2001) 228–233.
[101] V. Mayer-Schönberger, K. Cukier, Big Data: A Revolution That Will Transform How We Live, Work, and Think, John Murray Publishers, London, 2013.
[102] N. Metropolis, S. Ulam, The Monte Carlo method, Journal of the American Statistical Association 44 (1949) 335–341.
[103] N. Metropolis, A.W. Rosenbluth, M.N. Rosenbuth, A. Teller, H. Teller, Equations of state calculations by fast computing machines, Journal of Chemical Physics 21 (1953) 1087–1091.
[104] M. Minar, J. Naher, Recent advances in deep learning: an overview, arXiv:1807.08169, 2018. (Accessed 15 December 2018).
[105] T.M. Mitchell, Machine Learning, McGraw–Hill, New York, 1997.
[106] R.M. Neal, Bayesian Learning for Neural Networks, Springer-Verlag, Berlin, 1996.
[107] R.M. Neal, Regression and classification using Gaussian process priors, in: J.M. Bernardo, J.O. Berger, A.P. Dawid, A.F.M. Smith (Eds.), Bayesian Statistics, vol. 6, Oxford University Press, Oxford, 1998, pp. 475–501.
[108] R.M. Neal, Bayesian Methods for Machine Learning, NIPS Tutorial, University of Toronto, 2004.
[109] J.A. Nelder, R. Mead, A simplex method for function optimization, Computer Journal 7 (1965) 308–313.
[110] Y. Nesterov, A method for solving a convex programming problem with convergence rate of $O(1/k^2)$, Soviet Mathematics. Doklady 27 (2) (1983) 372–376.
[111] J. Nocedal, S.J. Wright, Numerical Optimization, 2nd edition, Springer, New York, 2006.
[112] J. Pearl, Probabilistic Reasoning in Intelligent Systems, Morgan Kaufmann, San Francisco, 1988.
[113] W. Pedrycz, F. Gomide, An Introduction to Fuzzy Sets: Analysis and Design, The MIT Press, Cambridge, MA, 1998.
[114] L.Y. Pratt, Discriminability-based transfer between neural networks, in: NIPS Conference: Advances in Neural Information Processing Systems, Morgan Kaufmann, Burlington, MA, 1993, pp. 204–211.
[115] W.H. Press, S.A. Teukolsky, W.T. Vetterling, B.P. Flannery, Numerical Recipes: The Art of Scientific Computing, 3rd edition, Cambridge University Press, Cambridge, 2007.

[116] N. Qian, On the momentum term in gradient descent learning algorithms, Neural Networks: Journal of International Neural Network Society 12 (1) (1999) 145–151.
[117] J.R. Quinlan, Induction of decision trees, Machine Learning 1 (1) (1986) 81–106.
[118] J.R. Quinland, C4.5: Programs for Machine Learning, Morgan Kaufmann, San Mateo, CA, 1993.
[119] R. Raina, A. Battle, H. Lee, B. Packer, A.Y. Ng, Self-taught learning: transfer learning from unlabelled data, in: Proceedings of 24th International Conference on Machine Learning, ICML, Corvallis, OR, 20–24 June, 2007, 2007, pp. 759–766.
[120] P. Ranganathan, C.S. Pramesh, R. Aggarwal, Common pitfalls in statistical analysis: logistic regression, Perspectives in Clinical Research 8 (3) (2017) 148–151.
[121] C.E. Rasmussen, C.K.I. Williamns, Gaussian Processes for Machine Learning, MIT Press, Cambridge, MA, 2006.
[122] L.S. Riza, C. Bergmeir, F. Herrera, J.M. Benitez, Frbs: fuzzy rule-based systems for classification and regression in R, Journal of Statistical Software 65 (6) (2015) 1–30.
[123] L. Rokach, O. Maimon, Decision trees, in: O. Maimon, L. Rokach (Eds.), Data Mining and Knowledge Discovery Handbook, Springer, Boston, MA, 2005, pp. 165–192.
[124] F. Rosenblatt, The perceptron: a probabilistic model for information storage and organization in the brain, Psychological Review 65 (6) (1958) 386–408.
[125] S. Ruder, An overview of gradient descent optimization algorithms, arXiv:1609.04747, September 2016, https://arxiv.org/abs/1609.04747. (Accessed 10 August 2018).
[126] D.E. Rumelhart, G.E. Hinton, R.J. Williams, Learning representations by back-propagating errors, Nature 323 (6088) (1986) 533–536.
[127] S. Russell, P. Norvig, Artificial Intelligence: A Modern Approach, 2nd edition, Prentice Hall, 2001.
[128] S. Sabour, N. Frosst, G.E. Hinton, Dynamic routing between capsules, in: 31st Conference on Neural Information Processing Systems, NIPS 2017, Long Beach, CA, arXiv: 1710.09829, 2017. (Accessed 20 December 2018).
[129] R. Shadmehr, Biological Learning and Control, MIT Press, Cambridge, MA, 2012.
[130] J. Shlens, A tutorial on principal component analysis, (2014) 1-12, https://arxiv.org/abs/1404.110, Version 3. (Accessed 20 July 2018).
[131] J. Shlens, A tutorial on independent component analysis, (2014) 1-13, https://arxiv.org/abs/1404.2986, Version 1. (Accessed 27 August 2018).
[132] L.I. Smith, A Tutorial on Principal Components Analysis, Technical Report OUCS-2002-12, University of Otago, New Zealand, 2002.
[133] A.J. Smola, B. Schölkopf, A tutorial on support vector regression, Statistics and Computing 14 (3) (2004) 199–222.
[134] P. Smolensky, Information processing in dynamical systems: foundations of harmony theory, Chapter 6, in: D.R. Rumelhart (Ed.), Parallel Distributed Processing: Explorations in the Microstructure of Cognition, vol. 1: Foundation, MIT Press, Cambridge, MA, 1986, pp. 194–281.
[135] N. Srivastava, G.E. Hinton, A. Krizhevsky, I. Sutskever, R. Salakhutdinov, Dropout: a simple way to prevent neural networks from overfitting, Journal of Machine Learning Research 15 (2014) 1929–1958.
[136] S.M. Stigler, Gauss and the invention of least squares, The Annals of Statistics 9 (3) (1981) 465–474.
[137] R. Storn, K. Price, Differential evolution: a simple and efficient heuristic for global optimization over continuous spaces, Journal of Global Optimization 11 (4) (1997) 341–359.
[138] P.N. Tan, M. Steibach, A. Karpatne, V. Kumar, Introduction to Data Mining, second edition, Pearson Education, London, 2019.

[139] R. Tibshirani, Regression shrinkage and selection via the Lasso, Journal of the Royal Statistical Society, Series B, Methodological 58 (1) (1996) 267–288.
[140] L.N. Trefethen, D. Bau III, Numerical Linear Algebra, Society for Industrial and Applied Mathematics, Philadelphia, 1997.
[141] L. Rorrey, J. Shavlik, Transfer learning, in: E. Soria, J. Martin, R. Magdalena, M. Martinez, A. Serrano (Eds.), Handbook of Research on Machine Learning Applications, IGI Global, Penn, USA, 2009, pp. 1–22.
[142] F. van Veen, The neural network zoo, Asimov Institute, online article, 14 Sept 2016. (Accessed 15 December 2018).
[143] V.N. Vapnik, The Nature of Statistical Learning Theory, Springer-Verlag, Berlin, 1995.
[144] P.H. Winston, Artificial Intelligence, 3rd edition, Addison-Wesley Publishing, Reading, MA, 1992.
[145] I.H. Witten, E. Frank, M.A. Hall, C.J. Pal, Data Mining: Practical Machine Learning Tools and Techniques, 4th edition, Morgan Kaufmann, Cambridge, MA, 2016.
[146] J. Wolberg, Data Analysis Using the Method of Least Squares: Extracting the Most Information From Experiments, Springer, Berlin, 2005.
[147] X.D. Wu, V. Kumar, J. Ross Quinlan, J. Ghosh, Q. Yang, H. Motoda, G.J. McLachlan, A. Ng, B. Liu, P.S. Yu, Z.H. Zhou, M. Steinbach, D.J. Hand, D. Steinberg, Top 10 algorithms in data mining, Knowledge and Information Systems 14 (1) (2008) 1–37.
[148] X.S. Yang, Nature-Inspired Metaheuristic Algorithms, Luniver Press, Bristol, UK, 2008.
[149] X.S. Yang, Firefly algorithm, stochastic test functions and design optimisation, International Journal of Bio-Inspired Computation 2 (2) (2010) 78–84.
[150] X.S. Yang, A new metaheuristic bat-inspired algorithm, in: J.R. González, D.A. Pelta, C. Cruz, G. Terraza, N. Krasnogor (Eds.), Nature-Inspired Cooperative Strategies for Optimization, NICSO 2010, in: SCI, vol. 284, Springer, Heidelberg, 2010, pp. 65–74.
[151] X.S. Yang, Engineering Optimization: An Introduction With Metaheuristic Applications, John Wiley and Sons, Hoboken, NJ, 2010.
[152] X.S. Yang, Bat algorithm for multi-objective optimisation, International Journal of Bio-Inspired Computation 3 (5) (2011) 267–274.
[153] X.S. Yang, Flower pollination algorithm for global optimization, in: J. Durand-Lose, N. Jonoska (Eds.), Unconventional Computation and Natural Computation, in: Lecture Notes in Computer Science, vol. 7445, 2012, pp. 240–249.
[154] X.S. Yang, S. Deb, Cuckoo search via Lévy flights, in: Proc. of World Congress on Nature & Biologically Inspired Computing, NaBic 2009, Coimbatore, India, IEEE Publications, USA, 2009, pp. 210–214.
[155] X.S. Yang, S. Deb, Engineering optimization by cuckoo search, International Journal of Mathematical Modelling and Numerical Optimisation 1 (4) (2010) 330–343.
[156] X.S. Yang, S. Deb, Cuckoo search: recent advances and applications, Neural Computing & Applications 24 (1) (2014) 169–174.
[157] X.S. Yang, M. Karamanoglu, X.S. He, Flower pollination algorithm: a novel approach for multiobjective optimization, Engineering Optimization 46 (9) (2014) 1222–1237.
[158] X.S. Yang, Cuckoo Search and Firefly Algorithm: Theory and Applications, Studies in Computational Intelligence, vol. 516, Springer, Heidelberg, 2014.
[159] X.S. Yang, Nature-Inspired Optimization Algorithms, Elsevier Insight, London, 2014.
[160] X.S. Yang, Engineering Mathematics With Examples and Applications, Academic Press, London, 2017.
[161] X.S. Yang, Optimization Techniques and Applications With Examples, John Wiley & Sons, Hoboken, NJ, 2018.
[162] L.A. Zadeh, Fuzzy sets, Information and Control 8 (3) (1965) 338–353.

[163] M.J. Zaki, W. Meira Jr., Data Mining and Analysis: Fundamental Concepts and Algorithms, Cambridge University Press, Cambridge, 2014.
[164] M.D. Zeiler, AdaDelta: an adaptive learning rate method, preprint, arXiv:1212.5701, 2012. (Accessed 10 August 2018).
[165] H. Zou, T. Hastie, Regularization and variable selection via the elastic net, Journal of the Royal Statistical Society, Series B 67 (2) (2005) 301–320.

Index